NATIONAL STANDARD OF THE PEOPLE'S REPUBLIC OF CHINA

Code for Construction of Steel Structures

GB 50755 - 2012

Chief Development Department: Ministry of Housing and Urban-Rural Development
of the People's Republic of China
Approval Department: Ministry of Housing and Urban-Rural Development
of the People's Republic of China
Implementation Date: August 1, 2012

China Architecture & Building Press

Beijing 2014

图书在版编目(CIP)数据

钢结构工程施工规范 GB 50755-2012/中华人民共和国住房和城乡建设部组织编译. —北京：中国建筑工业出版社，2014.9
（工程建设标准英文版）
ISBN 978-7-112-16937-5

Ⅰ.①钢… Ⅱ.①中… Ⅲ.①钢结构-工程施工-建筑规范-中国-英文 Ⅳ.①TU758.11-65

中国版本图书馆 CIP 数据核字(2014)第 116764 号

Chinese edition first published in the People's Republic of China in 2012
English edition first published in the People's Republic of China in 2014
by China Architecture & Building Press
No. 9 Sanlihe Road
Beijing, 100037
www.cabp.com.cn

Printed in China by BeiJing YanLinJiZhao printing CO., LTD

© 2012 by Ministry of Housing and Urban-Rural Development of
the People's Republic of China

All rights reserved. No part of this publication may be reproduced or transmitted in any form or
by any means, graphic, electronic, or mechanical, including photocopying, recording,
or any information storage and retrieval systems, without written permission of the publisher.

This book is sold subject to the condition that it shall not, by way of trade or otherwise, be lent,
re-sold, hired out or otherwise circulated without the publisher's prior consent in any form of
blinding or cover other than that in which this is published and without a similar condition
including this condition being imposed on the subsequent purchaser.

ISBN 978-7-112-16937-5(25728)

Announcement of the Ministry of Housing and Urban-Rural Development of the People's Republic of China

No. 1263

Announcement on Publishing the National Standard
Code for Construction of Steel Structures

Code for Construction of Steel Structures has been approved as a national standard with a serial number of GB 50755 - 2012 and shall be implemented from August 1, 2012. Thereinto, Articles 11.2.4 and 11.2.6 are compulsory provisions and must be enforced strictly.

Authorized by the Standard Rating Research Institute of the Ministry of Housing and Urban-Rural Development of the People's Republic of China, this Code is published by China Architecture & Building Press.

Ministry of Housing and Urban-Rural Development of the People's Republic of China
January 21, 2012

Foreword

This Code was formulated by China State Construction Engineering Corporation Limited and China Construction Steel Structure Corp. Ltd. jointly with organizations concerned according to the requirements of Document JIANBIAO [2007] NO. 125 issued by Ministry of Housing and Urban-Rural Development of the People's Republic of China (MOHURD)-"Notice on Printing the Development and Revision Plan (the First Batch) of National Engineering Construction Standards and Codes in 2007".

This Code is a general technical standard for the construction of steel structures; it provides basic requirements for the construction of steel structures and its process control and is regarded as the basis to formulate and revise relevant specialized standards. During the process of formulating this Code, the drafting group conducted extensive investigation, summarized the experiences in the construction of steel structures over decades, referred to relevant foreign standards, carried out monographic studies, extensively asked for the opinions of organizations concerned and experts in various ways, conducted repeated discussion, coordination and modification on the major issues, and finalized this Code through review.

This Code comprises 16 chapters, including general provisions, terms and symbols, basic requirements, design of construction stage, materials, welding, connecting of fasteners, processing of steel parts and components, assembling and processing of members, test assembling of steel structures, installation of steel structures, profiled metal plate, coating, construction survey, construction monitoring, construction safety and environmental protection.

The provisions printed in bold type in this Code are compulsory ones and must be enforced strictly.

Ministry of Housing and Urban-Rural Development (MOHURD) is in charge of the administration of this Code and the explanation of the compulsory provisions, and China State Construction Engineering Corporation Limited is responsible for the explanation of specific technical contents. With a view to improving the quality of this Code, all relevant organizations are kindly requested to sum up and accumulate your experiences in actual practices during the process of implementing this Code. The relevant opinions and advice, whenever necessary, can be posted or passed on to the Science and Technology Department of China State Construction Engineering Corporation Limited in Central Building for future reference in revision (Address: No. 15 Shanlihe Road, Xicheng District, 100037, Beijing, China; E-mail: gb50755@cscec.com.cn).

Chief development organization of this code:
    China State Construction Engineering Corporation Limited
    China Construction Steel Structure Corp. Ltd.

Participating development organizations of this code:
    China Construction Third Engineering Bureau Co., Ltd.
    Shanghai Mechanized Construction Corporation Ltd.
    Zhejiang Southeast Space Frame Company

BAOSTEEL Construction Co., Ltd.
Central Research Institute of Building and Construction of MCC Group Co., Ltd.
Jiangsu Huning Steel Mechanism Co., Ltd.
China Northeast Architecture Design and Research Institute Co., Ltd.
Shanghai Construction Group Co., Ltd.
China Construction Second Engineering Bureau Co., Ltd.
China Construction Industrial Equipment Installation Co., Ltd.
Beijing Building Construction Research Institute Co., Ltd.
Hempel (China) Co., Ltd.
China Construction Steel Structure Jiangsu Co., Ltd.
China Jingye Engineering Corporation Limited

Chief drafting staff of this Code:

Mao Zhibing　Zhang Kun　Xiao Xuwen　Wang Hong
Dai Lixian　Chen Zhenming　Zhang Jingbo　Zhou Guangen
Wu Xinzhi　He Mingxuan　Hou Zhaoxin　Lu Kekuan
Bao Guangjian　Fei Xinhua　Chen Xiaoming　Liao Gonghua
Pang Jinghui　Sun Zhe　Fang Jun　Ma Hesheng
Wu Julong　Qin Jie　Wu Haobo　Cui Xiaoqiang
Liu Shimin　Bian Ruoning　Li Xiaoming

Chief examiners of this Code:

Ma Kejian　Chen Luru　Wang Dasui　He Xianjuan
Yang Sixin　Jin Hugen　Chai Chang　Fan Maoda
Guo Yanlin　Wang Cuikun　Shu Weinong

Contents

1 General Provisions ··· (1)
2 Terms and Symbols ··· (2)
   2.1 Terms ··· (2)
   2.2 Symbols ·· (3)
3 Basic Requirements ··· (4)
4 Design of Construction Stage ·· (5)
   4.1 General Requirements ··· (5)
   4.2 Structure Analysis of Construction Stage ·· (5)
   4.3 Preset Deformation of Structure ·· (6)
   4.4 Detail Design ··· (6)
5 Materials ·· (8)
   5.1 General Requirements ··· (8)
   5.2 Steel Materials ··· (8)
   5.3 Welding Materials ··· (10)
   5.4 Fasteners for Connections ··· (11)
   5.5 Steel Castings, Anchorages and Pins ·· (12)
   5.6 Coating Materials ··· (12)
   5.7 Material Storage ··· (12)
6 Welding ·· (14)
   6.1 General Requirements ··· (14)
   6.2 Welding Personnel ··· (14)
   6.3 Welding Procedure ··· (14)
   6.4 Welded Joints ·· (18)
   6.5 Welding Quality Inspection ·· (20)
   6.6 Repair of Welding Defects ··· (20)
7 Connecting of Fasteners ··· (21)
   7.1 General Requirements ··· (21)
   7.2 Preparation of Connecting Pieces and Friction Surface Treatment ········ (21)
   7.3 Connection of Ordinary Fasteners ·· (22)
   7.4 Connection of High Strength Bolts ··· (23)
8 Processing of Steel Parts and Components ·· (27)
   8.1 General Requirements ··· (27)
   8.2 Setting out and Marking-off ·· (27)
   8.3 Cutting ··· (27)
   8.4 Rectification and Formation ·· (28)
   8.5 Trimming of Edges ·· (30)
   8.6 Holing ·· (31)

  8.7 Processing of Bolted Spheres and Welded Hollow Spheres .... (31)
  8.8 Processing of Cast Steel Node .... (33)
  8.9 Processing of Rope Nodes .... (33)
9 Assembling and Processing of Members .... (34)
  9.1 General Requirements .... (34)
  9.2 Assembling of Components .... (34)
  9.3 Assembling of Members .... (35)
  9.4 Milling of Member Ends .... (36)
  9.5 Rectification of Members .... (36)
10 Test Assembling of Steel Structures .... (37)
  10.1 General Requirements .... (37)
  10.2 Test Assembling of Steel Structures .... (37)
  10.3 Test Assembling of Computer Assistance Simulation .... (38)
11 Installation of Steel Structures .... (39)
  11.1 General Requirements .... (39)
  11.2 Hoisting Equipment and Hoisting Mechanism .... (39)
  11.3 Foundation, Bearing Surface and Embedment Parts .... (40)
  11.4 Installation of Members .... (41)
  11.5 Single-story Steel Structures .... (43)
  11.6 Multi-story and Tall Steel Structures .... (43)
  11.7 Long-span Spatial Steel Structures .... (44)
  11.8 High-rising Steel Structures .... (45)
12 Profiled Metal Plate .... (46)
13 Coating .... (47)
  13.1 General Requirements .... (47)
  13.2 Surface Treatment .... (47)
  13.3 Coating of Anticorrosive Paint Layer .... (48)
  13.4 Metal Hot Spraying .... (49)
  13.5 Anticorrosion of Hot Dipping Galvanizing .... (49)
  13.6 Coating of Fire-retardant Coating Layer .... (50)
14 Construction Survey .... (52)
  14.1 General Requirements .... (52)
  14.2 Plan Control Network Survey .... (52)
  14.3 Elevation Control Network Survey .... (53)
  14.4 Survey of Single-story Steel Structures .... (53)
  14.5 Survey of Multi-story and Tall Steel Structures .... (54)
  14.6 Survey of High-rising Steel Structures .... (54)
15 Construction Monitoring .... (56)
  15.1 General Requirements .... (56)
  15.2 Construction Monitoring .... (56)
16 Construction Safety and Environmental Protection .... (58)
  16.1 General Requirements .... (58)

16.2	Climb Up	(58)
16.3	Safety Channel	(58)
16.4	Protection of Portals and Sides	(59)
16.5	Construction Machinery and Equipment	(59)
16.6	Safety of Site Hoisting Area	(59)
16.7	Fire Safety Measures	(60)
16.8	Environmental Protection Measures	(60)

Explanation of Wording in This Code ····· (61)
List of Quoted Standards ····· (62)

1 General Provisions

1.0.1 This Code is formulated with a view to implementing the national technical and economic policies in the construction of steel structures and achieving safety and usability, guaranteed quality, advanced technology, and economy and rationality.

1.0.2 This Code is applicable to construction of steel structures in industrial and civil buildings and structures.

1.0.3 The construction of steel structures shall be carried out according to those specified in this Code and quality acceptance shall be carried out for it according to the current national standards GB 50300 *Unified Standard for Constructional Quality Acceptance of Building Engineering* and GB 50205 *Code for Acceptance of Construction Quality of Steel Structures*.

1.0.4 The construction of steel structures shall not only meet the requirements of this Code, but also comply with those in the current relevant standards of the nation.

2 Terms and Symbols

2.1 Terms

2.1.1 Design document

A general term for such technical documents as design drawing, design instruction and design change completed by the design organization.

2.1.2 Design drawing

The technical drawing prepared by the design organization as the engineering construction basis.

2.1.3 Detail drawing for construction

The detailed technical drawing drawn according to the technical requirements of the steel structure design drawing and construction processes for direct guidance on fabrication and installation of steel structures.

2.1.4 Temporary structure

The structure that is erected during the construction period and required to be removed at the end of the construction.

2.1.5 Temporary measure

Some necessary structures or temporary parts and member bars, such as lifting hole, connecting plate and auxiliary member that are arranged with a view to meeting the construction demands and guaranteeing engineering safety during the construction period.

2.1.6 Space rigid unit

The basically-stable space system that is composed of members.

2.1.7 Welded hollow spherical node

The node of the pipe that is directly welded to the sphere.

2.1.8 Bolted spherical node

The node that connects the pipe and sphere via -bolts and is composed of parts and components such as the bolted sphere, high strength bolts, sleeve, fastening screw and conical head or closing plate.

2.1.9 Mean slip coefficient

The ratio of the external sliding force to the normal pressure of the connection when faying surface sliding.

2.1.10 Structure analysis of construction stage

The structure analysis and calculation carried out in order to meet the requirements of relevant functions during the fabrication, transportation and installation of steel structures.

2.1.11 Preset deformation

The initial deformation set in advance in order to make the structure or member reach the control objectives of the designed geometric positioning after the completion of the construction.

2.1.12 Test assembling

The test assembling carried out in advance to inspect whether the shape and dimension of the member meet the quality requirements.

2.1.13 Ambient temperature

The temperature at the site during the fabrication or installation

2.2 Symbols

2.2.1 Geometric parameters

b ——width or free outstanding width of the plate;
d ——diameter;
f ——deflection and bending rise;
h ——section height;
l ——length or span;
m ——nominal thickness of high strength nut;
n ——number of washers;
r ——radius;
s ——normal thickness of high strength washer;
t ——plate and wall thickness;
p ——thread pitch;
Δ ——gap between contact surfaces or increment;
H ——column height;
R_a ——surface roughness parameter.

2.2.2 Action and load

P ——design pretension force of the high strength bolt;
T ——torque of the high strength bolt.

2.2.3 Other

k ——coefficient.

3 Basic Requirements

3.0.1 The construction organization of steel structures shall have corresponding construction qualification as well as safety, quality and environmental management system.

3.0.2 Before the construction of the steel structures, technical documents such as construction organization design and its supporting specific construction plan approved by the technical principal of the construction organization shall be provided and submitted to the supervision engineer or the Owner's representative according to the relevant requirements; experts shall be organized to review the technical plan and safety emergency plan for the construction of important steel structures.

3.0.3 In the technical documents and technical contract documents with regard to the construction of steel structures, the construction quality shall at least meet the relevant requirements of the current national standard GB 50205 *Code for Acceptance of Construction Quality of Steel Structures*.

3.0.4 Fabrication and installation of steel structures shall meet the requirements of the design drawing. The construction organization shall carry out process review on the design documents; where it is necessary to change the design, consent from the original design organization shall be obtained and relevant design change documents shall be transacted.

3.0.5 During the construction and quality acceptance of steel structures, effective measuring instruments & tools shall be used. The construction organizations and supervision organizations in various professions shall unify the measurement standard.

3.0.6 Special equipment and tool for construction of steel structures shall meet the construction requirements and shall be within the validity of the qualified verification.

3.0.7 For the construction of steel structures, quality process control shall be carried out according to the following requirements:

    1 Site acceptance is carried out for the materials and finished products; re-inspection is carried out for the materials and semi-finished products related to the safety and function according to the relevant requirements; witness sampling and sample delivery are adopted;

    2 Quality control is carried out for each process according to the requirements of the construction processes and process inspection is carried out;

    3 Handing-over inspection is carried out between relevant professions and work types;

    4 Quality acceptance is carried out before the closure of the concealed work.

3.0.8 The new technology, process, material and structure not involved in this Code shall be tested before use for the first time, for which necessary supplementary standards shall be determined according to the test results and subject to the expert demonstration.

4 Design of Construction Stage

4.1 General Requirements

4.1.1 This Chapter is applicable to the design of steel structure construction stage, including structure analysis and checking, design for the preset structure deformation and detail drawing design.

4.1.2 During the design of construction stage, the selected design index shall meet the relevant requirements of the design documents and the current national standard GB 50017 *Code for Design of Steel Structures*.

4.1.3 During the structure analysis and checking at the construction stage, loads shall meet the following requirements:

　　1 Dead load shall include structure self-weight and prestress and its characteristic value shall be calculated according to actual conditions;

　　2 Live load during the construction shall include construction loading and weight of operators and small tools; and its characteristic value may be calculated according to actual conditions;

　　3 Wind load may be taken according to the construction site and actual construction conditions, on the basis of at least 10-year mean recurrence interval; the wind load shall be calculated according to the relevant requirements of the current national standard GB 50009 *Load Code for the Design of Building Structures*. In case of any wind load greater than the ten-year wind pressure during the construction period, an emergency plan shall be worked out;

　　4 Snow load shall be taken and calculated according to the relevant requirements of the current national standard GB 50009 *Load Code for the Design of Building Structures*;

　　5 Ice load shall be taken and calculated according to the relevant requirements of the current national standard GB 50135 *Code for Design of High-rising Structures*;

　　6 Characteristic values of hoisting equipment and other equipment loads should be taken according to the equipment instruction;

　　7 The temperature action should be calculated according to the temperature difference change in the local meteorological information; the temperature difference on the sunny side and dark side of the structure due to sunlight should be calculated according to the relevant requirements of the current national standard GB 50135 *Code for Design of High-rising Structures*;

　　8 Load and action not specified in Items 1~7 of this article may be determined according to specific engineering conditions.

4.2 Structure Analysis of Construction Stage

4.2.1 When the construction method or sequence of steel structures has larger influence on their internal force and deformation or special requirements are made in the design documents, the structure analysis of construction stage shall be carried out; their strength, stability and stiffness during the construction stage shall be checked and the results shall meet the design requirements.

4.2.2 Load effect combination and partial load coefficient for the structure analysis of construction stage shall meet the requirements of the current national standard GB 50009 *Load Code for the Design of Building Structures*.

4.2.3 In the structure analysis of construction stage, the structure importance coefficient shall be not less than 0.9 and the importance coefficient of the important temporary structure shall be not less than 1.0.

4.2.4 Load action in the construction stage, structure analysis model and basic assumption in the construction stage shall be corresponding with the actual construction conditions. For the structure in the construction stage, elastic analysis should be carried out using statics methods.

4.2.5 For the temporary structure and measures in the construction stage, member strength, stability and stiffness shall be checked according to the load action during the construction; the strength and stability of the connections shall be checked. When the temporary structure serves as a bearing structure for the equipment, special design shall be carried out; when the temporary structure or measure has relatively large influence on the structure, it shall be submitted to the original design organization for confirmation.

4.2.6 Removal sequence and procedure of the temporary structure shall be determined through analysis and calculation. Specific construction plan shall be prepared; where necessary, it shall be subject to the expert demonstration.

4.2.7 For the member or structure unit under hoisting conditions, the strength, stability and deformation should be checked and $1.1 \sim 1.4$ should be taken for the dynamic factor.

4.2.8 Installation and tension sequence of the cables in the cable structure shall be determined through analysis and calculation and specific construction plan shall be prepared; the calculation results shall be confirmed by the original design organization.

4.2.9 For the ground or floor that bears the mobile hoisting equipment, the bearing capacity and deformation shall be checked. When the supporting ground is at or adjacent to the side slope, stability of the side slope shall be checked.

4.3 Preset Deformation of Structure

4.3.1 During the normal use or in the construction stage, when any deformation due to the deadweight and other load action exceeds the limit specified in the design documents or current relevant standards of the nation or preset deformation requirements are made for the major structure in the design documents, preset deformation shall be adopted for the structure during the construction period.

4.3.2 When the preset deformation of structure is calculated, the characteristic value shall be taken for the load; the load effect combination shall meet the relevant requirements of the current national standard GB 50009 *Load Code for the Design of Building Structures*.

4.3.3 The preset deformation of structure shall be calculated through the structure analysis and in combination with the construction process; it shall be jointly determined by the construction organization and the original design organization. Specific process design shall be carried out for the preset deformation of structure.

4.4 Detail Design

4.4.1 The detail drawing for construction of steel structures shall be prepared according to structure design documents and relevant technical documents and shall be confirmed by the original design organization; where the connection design is necessary, the connection design document also shall be confirmed by the original design organization.

4.4.2 The detail design of steel structures shall meet relevant technical requirements such as construction structures, construction process and member transportation.

4.4.3 The detail drawing for construction of steel structures shall include drawing list, general design description, member layout, detail drawing for the member and connection installation etc.. The drawing shall be shown clearly and completely; for the detail drawing for construction of complex-space member and connection/node, three dimensional diagram should be drawn.

4.4.4 The member weight shall be calculated and listed in the detail drawing for construction of steel structures; the weight of steel plate parts should be calculated as a rectangle and the weld weight should be calculated using 1.5% weight of the welded member.

5 Materials

5.1 General Requirements

5.1.1 This Chapter is applicable to ordering, site acceptance, re-inspection and storage management of steel structure materials.

5.1.2 Materials used in the steel structures shall meet the requirements of design documents and current relevant standards of the nation; they shall be provided with quality certificates and used after the qualification of the site inspection.

5.1.3 The construction organization shall establish material management system and achieve normalized ordering, storage and use.

5.2 Steel Materials

5.2.1 When the steel materials are ordered, the variety, specification and performance shall meet the requirements of the design documents and current standards of the nation with regard to steel materials; the standards of common steel products should be adopted according to those specified in Table 5.2.1.

Table 5.2.1 Standards of Common Steel Products

Serial numbers	Standard names
GB/T 699	Quality Carbon Structural Steels
GB/T 700	Carbon Structural Steels
GB/T 1591	High Strength Low Alloy Structural Steels
GB/T 3077	Alloy Structure Steels
GB/T 4171	Atmospheric Corrosion Resisting Structural Steel
GB/T 5313	SteelPlates with Through-thickness Characteristics
GB/T 19879	Steel Plates for Building Structure
GB/T 247	General Rule of Package Mark and Certification for Steel Plates (Sheets) and Strips
GB/T 708	Dimension Shape Weight and Tolerance for Cold-Rolled Steel Plates and Sheets
GB/T 709	Dimension Shape Weight and Tolerances for Hot-Rolled Steel Plates and Sheets
GB 912	Hot-Rolled Sheets and Strips of Carbon Structural Steels and High Strength Low Alloy Structural Steels
GB/T 3274	Hot-Rolled Plates and Strips of Carbon Structural Steels and High Strength Low Alloy Structural Steels
GB/T 14977	General Requirement for Surface Condition of Hot-Rolled Steel Plates
GB/T 17505	Steel and Steel Products-General Technical Delivery Requirements
GB/T 2101	General Requirement of Acceptance Packaging Marking and Certification for Section Steel
GB/T 11263	Hot-Rolled H and Cut T Section Steel
GB/T 706	Hot Rolled Section Steel
GB/T 8162	Seamless Steel Tubes for Structural Purposes
GB/T 13793	Steel Pipes with a Longitudinal Electric (Resistance) Weld
GB/T 17395	Dimensions, Shapes, Masses and Tolerances of Seamless Steel Tubes
GB/T 6728	Cold Formed Steel Hollow Sections For General Structure-Dimensions, Shapes, Weight and Permissible Deviations
GB/T 12755	Profiled Steel Sheet for Building
GB 8918	Steel Wire Ropes for Important Purposes
YB 3301	The Welded Steel H-Section
YB/T 152	High Strength Low Relaxation Hot-dip Galvanized Steel Strand for Prestress
YB/T 5004	Zinc-coated Steel Wire Strands
GB/T 5224	Steel Strand for Prestressed Concrete
GB/T 17101	Hot-dip Galvanized Steel Wires for Bridge Cables
GB/T 20934	Steel Tie Rod

5.2.2 The steel material order contract shall clearly specify the designation, specification/dimension, performance index, inspection requirements and dimension tolerance. Re-inspection sampling allowance shall be reserved for the steel materials of a scale; the delivery conditions of the steel materials should be determined through negotiation with the Supplier according to the performance requirements for the steel materials specified in the design document.

5.2.3 The site acceptance for the steel materials shall not only meet those specified in this Code, but also meet the relevant requirements of the current national standard GB 50205 *Code for Acceptance of Construction Quality of Steel Structures*. For steel materials under one of the following conditions, sampling re-inspection shall be carried out:

    1 Imported steel materials;

    2 Mixed lot;

    3 The plate thickness is greater than or equal to 40mm and they are designed with thick plates in accordance with Z-direction performance;

    4 The safety grade of building structure is Grade I; the steel materials adopted for the primary load-carrying member in the large-span steel structure;

    5 The steel materials with design re-inspection requirements;

    6 Steel materials with doubtful quality.

5.2.4 The re-inspection for the steel materials shall cover the mechanical performance test and chemical composition analysis; the sampling, sample preparation and tests may be carried out according to the standards listed in Table 5.2.4.

Table 5.2.4 Steel Test Standards

Serial numbers	Standard names
GB/T 2975	Steel and Steel Products-Location and Preparation of Test Pieces for Mechanical Testing
GB/T 228.1	Metallic Materials-Tensile Testing-Part 1: Method of Test at Room Temperature
GB/T 229	Metallic Materials-Charpy Pendulum Impact Test Method
GB/T 232	Metallic Materials-Bend Test
GB/T 20066	Steel and Iron-Sampling and Preparation of Samples for the Determination of Chemical Composition
GB/T 222	Permissible Tolerances for Chemical Composition of Steel Products
GB/T 223	Methods for Chemical Analysis of Iron, Steel and Alloy

5.2.5 When no special requirements are made in the design document, the sampling re-inspection lot of the steel materials with common designation should be in accordance with the following requirements:

    1 For Q235 and Q345 steel materials with the plate thickness less than 40mm, the steel materials of the same manufacturer, designation and quality grade shall form an inspection lot with the weight not greater than 150t; when the supply weight of the steel materials of the same manufacturer and designation is greater than 600t and all the re-inspections are qualified, the weight of each lot may be increased to 400t;

    2 For Q235 and Q345 steel materials with the plate thickness greater than or equal to 40mm, the steel materials of the same manufacturer, designation and quality grade shall form an inspection lot with the weight not greater than 60t; when the supply weight of the steel materials of the same manufacturer and designation is greater than 600t and all the re-inspections are qualified, the weight of each lot may be increased to 400t;

3 For Q390 steel materials, the steel materials of the same manufacturer and quality grade shall form an inspection lot with the weight not greater than 60t; when the supply weight of the steel materials of the same manufacturer is greater than 600t and all the re-inspections are qualified, the weight of each lot may be increased to 300t;

4 For Q235GJ, Q345GJ and Q390GJ steel materials, the steel materials of the same manufacturer, designation and quality grade shall form an inspection lot with the weight less than or equal to 60t; when the supply weight of the steel materials of the same manufacturer and designation is greater than 600t and all the re-inspections are qualified, the weight of each lot may be increased to 300t;

5 For Q420, Q460, Q420GJ and Q460GJ steel materials, each inspection lot shall be composed of the steel materials of the same designation, quality grade, batch Number, thickness and delivery condition and weight of each lot shall be less than or equal to 60t;

6 For those with thickness direction requirements, the steel plates should be re-inspected by ultrasonic nondestructive testing mode one by one.

5.2.6 The re-inspection sampling, sample preparation and test methods of imported steel materials shall be in accordance with the design documents and contract provisions. The commodity inspection results may be regarded as effective material re-inspection results if approved by the supervision engineer.

5.3 Welding Materials

5.3.1 The variety, specification and performance of welding materials shall meet the requirements of the current relevant product standards of the nation and the design requirements. The standards of common welding materials & products should be adopted according to those specified in Table 5.3.1. Welding materials such as electrode, welding wire, welding flux and consumable nozzle for electroslag welding shall be matched with the steel materials selected in the design and shall meet the relevant requirements of the current national standard GB 50661 *Code for Welding of Steel Structures*.

Table 5.3.1 Standards of Common Welding Materials & Products

Serial numbers	Standard names
GB/T 5117	Carbon Steel Covered Electrodes
GB/T 5118	Low Alloy Steel Covered Electrodes
GB/T 14957	Steel Wires for Melt Welding
GB/T 8110	Welding Electrodes and Rods for Gas Shielding Arc Welding of Carbon and Low Alloy Steel
GB/T 10045	Carbon Steel Flux Cored Electrodes for Arc Welding
GB/T 17493	Low Alloy Steel Flux Cored Electrodes for Arc Welding
GB/T 5293	Carbon Steel Electrodes and Fluxs for Submerged Arc Welding
GB/T 12470	Low-alloy Steel Electrodes and Fluxes for Submerged Arc Welding
GB/T 10432.1	Unthreaded Studs for Drawn Arc Stud Welding with Ceramic Ferrule
GB/T 10433	Cheese Head Studs for Arc Stud Welding

5.3.2 Sampling re-inspection shall be carried out for the welding materials for the important weld or those with doubt on the quality certificate; during the re-inspection, one group of welding wires should be taken from five lots (batches) for test and one group of the electrode should be taken from three lots (batches) for test.

5.3.3 Welding and cutting gases shall meet the requirements of the current national standard GB 50661 *Code for Welding of Steel Structures* and the standards listed in Table 5.3.3.

Table 5.3.3 Standards of Common Welding and Cutting Gases

Serial numbers	Standard names
GB/T 4842	Argon
GB/T 6052	Industrial Liquid Carbon Dioxide
HG/T 2537	Carbon Dioxide for Welding Use
GB 16912	Safety Technical Regulation For Oxygen and Relative Gases Produced with Cryogenic Method
GB 6819	Dissolved Acetylene
HG/T 3661.1	Burning Gases for Welding and Cutting-Propene
HG/T 3661.2	Burning Gases for Welding and Cutting-Propane
GB/T 13097	Epichlorohydrin for Industrial Use
HG/T 3728	Mixed Gas for Welding-Argon-Carbon Dioxide

5.4 Fasteners for Connections

5.4.1 Fasteners such as ordinary bolts, set of high strength large hexagon head bolt and set of tor-shear type high strength bolt for connections of steel structures shall meet the requirements of the standards listed in Table 5.4.1.

Table 5.4.1 Standards of Fasteners for Connection of Steel Structures

Serial numbers	Standard names
GB/T 5780	Hexagon Head Bolts- Product Grade C
GB/T 5781	Hexagon Head Bolts-Full Thread-Product Grade C
GB/T 5782	Hexagon Head Bolts
GB/T 5783	Hexagon Head Bolts-Full Thread
GB/T 1228	High Strength Bolts with Large Hexagon Head for Steel Structures
GB/T 1229	High Strength Large Hexagon Nuts for Steel Structures
GB/T 1230	High Strength Plain Washers for Steel Structures
GB/T 1231	Specifications of High Strength Bolts With Large Hexagon Head, Large Hexagon Nuts, Plain Washers for Steel Structures
GB/T 3632	Sets of Torshear Type High Strength Bolt Hexagon Nut and Plain Washer for Steel Structures
GB/T 3098.1	Mechanical Properties of Fasteners-Bolts, Screws and Studs

5.4.2 The set of high strength large hexagon head bolt and the set of tor-shear type high strength bolt shall be provided with accompanying delivery inspection reports with qualified torque coefficient and fastening axial force (pretension force) respectively. When the set of high strength bolt is used after 6-month storage period, torque coefficient or fastening axial force test shall be carried out again according to relevant requirements and only the qualified one may be put into service.

5.4.3 Torque coefficient and fastening axial force (pretension force) re-inspection shall be carried out respectively for the set of high strength large hexagon head bolt and the set of tor-shear type high strength bolt. The test bolt shall be randomly taken from the bolt lot to be installed at the construction site. 8 sets of bolts shall be randomly taken from each lot for re-inspection.

5.4.4 The safety grade of the building structure is Grade I; for the steel space frame structure of bolted spherical node with a span of 40m or above, surface hardness test shall be carried out for its connected high strength bolts; the surface harnesses of Grades 8.8 and 10.9 high strength bolts shall be HRC 21~29 and HRC 32~36 respectively and they shall be free from any crack or damage.

5.4.5 When the ordinary bolt is regarded as a permanent connecting bolt and it is specified in the design document or the quality is doubtful, minimum tensile load re-inspection for the physical bolts shall be carried out; 8 bolts of the same specification shall be randomly inspected.

5.5 Steel Castings, Anchorages and Pins

5.5.1 Casting materials selected for the steel castings shall meet the requirements of the standards listed in Table 5.5.1 and the design document.

Table 5.5.1 **Standards of Steel Castings**

Serial numbers	Standard names
GB/T 11352	Carbon Steel Castings for General Engineering Purpose
GB/T 7659	Steel Casting Suitable for Welded Structure

5.5.2 The prestress steel structure anchorages shall be selected according to the variety, anchorage requirements and pretension process etc; the anchorage materials shall meet the relevant requirements of the design documents, the current national standard GB/T 14370 *Anchorage, Grip and Coupler for Prestressing Tendons* and the professional standard JGJ 85 *Technical Specification for Application of Anchorage, Grip and Coupler for Prestressing Tendons*.

5.5.3 The specification and performance of the pins shall meet the relevant requirements of the design document and the current national standard GB/T 882 *Clevis Pin with Head*.

5.6 Coating Materials

5.6.1 Anti-corrosive paints, diluent and curing agent for the steel structures shall be selected according to the requirements of the design document and the current relevant product standards of the nation; the variety, specification and performance shall meet the requirements of the design document and the current relevant product standards of the nation.

5.6.2 Zinc content of the zinc rich anti-corrosive paint shall meet the relevant requirements of the design document and the current professional standard HG/T 3668 *Zinc Rich Primer*.

5.6.3 Variety and technical performance of the fire-proof coating shall meet the relevant requirements of the design document and the current national standard GB 14907 *Fire Resistive Coating for Steel Structure*.

5.6.4 Construction quality acceptance of the fire-proof coating for steel structure shall meet the relevant requirements of the current national standard GB 50205 *Code for Acceptance of Construction Quality of Steel Structures*.

5.7 Material Storage

5.7.1 Specified person, who should take the enterprise training, shall be assigned to be in charge of the management of material storage and finished products.

5.7.2 Materials shall be inspected before the warehousing. The variety, specification, lot No., quality certificate, Chinese mark and inspection report of the materials shall be confirmed and the surface quality and packages shall be inspected.

5.7.3 Materials that are qualified in the inspection shall be stacked by class according to their variety, specification and lot No. Signs shall be provided for the stacking of materials.

5.7.4 Warehousing and distribution of the materials shall be recorded. At the distribution and collection, the variety, specification and performance of the materials shall be checked.

5.7.5 The surplus materials shall be recycled for management. When they are recycled and warehoused, their variety, specification and quantity shall be checked and they shall be stored by class.

5.7.6 The steel materials shall be stacked such that their deformation and rust are minimized and supported by wood pads or cushion blocks.

5.7.7 Storage of welding materials shall meet the following requirements:

 1 Welding materials such as electrodes, welding wires and welding flux shall be respectively stored in a dry storage chamber according to the variety, specification, lot No.

 2 Electrodes, welding flux and porcelain ring of the stud shall be baked according to the requirements of the product instruction prior to use.

5.7.8 Connecting fasteners shall be prevented from being rusted or bumped and those of different lots shall not be stored together.

5.7.9 The coating materials shall be stored following the requirements of product instruction.

6 Welding

6.1 General Requirements

6.1.1 This Chapter is applicable to Shielded Metal Arc Welding, Gas Metal Arc Welding, Submerged Arc Welding, Electroslag Welding and Stud Welding during the construction of steel structures.

6.1.2 The steel structures construction organization shall meet the basic conditions and personnel qualification of the current national standard GB 50661 *Code for Welding of Steel Structures*.

6.1.3 Welding symbols in the welding construction drawing shall meet the relevant requirements of the current national standards GB/T 324 *Weld Symbolic Representation on Drawings* and GB/T 50105 *Standard for Structural Drawings*. In the drawings, the factory and on-site weld position, welding type, groove type and weld dimension shall be indicated.

6.1.4 The dimension of the weld groove shall follow the relevant requirements of the current national standard GB 50661 *Code for Welding of Steel Structures* and the change of groove dimension shall be operated only after the procedure qualification is qualified.

6.2 Welding Personnel

6.2.1 Welding technician (welding engineer) shall be qualified by corresponding qualification certificate; for large and important steel structures, the welding technical principal shall obtain intermediate technical title or above and have over five-year welding production or construction experience.

6.2.2 The welding quality inspectors shall have received the welding technology training and obtain corresponding quality inspection qualification certificate after the post training.

6.2.3 Weld non-destructive testing personnel shall obtain a grade certificate issued by the national professional assessment institution and be engaged in weld non-destructive testing inspection according to the qualified items in the certificate and the authority.

6.2.4 The welder shall pass through the examination and obtain a qualification certificate; he/she shall carry out the welding within the approved range and is strictly forbidden to take up his/her post without a certificate.

6.3 Welding Procedure

I Welding Procedure Qualification and Scheme

6.3.1 For parameters such as the steel materials, welding materials, welding method, joint type, welding position and post-weld heat treatment as well as combination of parameters that are firstly adopted by the construction organization, the welding procedure qualification test shall be carried out before the fabrication and installation of steel structures. The welding procedure qualification test methods and requirements as well as the limit conditions of procedure qualification exemption shall meet the relevant requirements of the current national standard GB 50661 *Code for Welding* *of Steel Structures.*

6.3.2 Before the welding procedure, the construction organization shall prepare the welding procedure document on the basis of qualified welding procedure qualification results or adopting satisfactory procedure qualification-free conditions and it shall cover the following items:

    1 Welding method or combination of the welding methods;
    2 Specification, designation, thickness and coverage of the base material;
    3 Specification, type and model of the filler metal;
    4 Welded joint type, groove type, dimension and the permissible tolerance;
    5 Welding position;
    6 Type and polarity of welding source;
    7 Back gouging;
    8 Welding parameters (welding current, welding voltage, welding speed, welding layer and weld bead distribution);
    9 Preheating temperature and inter-pass temperature range;
    10 Post-weld stress-relief treatment process;
    11 Other necessary requirements.

II Welding Conditions

6.3.3 During the welding, the ambient temperature, relative humidity and wind speed in the operating area shall meet the following requirements; when they are out of the range specified here and the welding is necessary to be carried out, special plan shall be prepared:

    1 The ambient temperature for operation shall be not lower than $-10°C$;
    2 Relative humidity in the welding operation area shall be not greater than 90%;
    3 When the manual Arc Welding and Self-Shielded Flux-Cored Wire Arc Welding are carried out, the maximum wind speed in the welding operation area shall be not greater than 8m/s; when the gas shielded arc welding is carried out, the maximum wind speed in the welding operation area shall be not greater than 2m/s.

6.3.4 For the on-site overhead welding, stable operation platform and protection shed shall be erected.

6.3.5 Before welding, tools such as wire brush and grinding wheel shall be adopted to remove the foreign bodies such as scale, rust and oil stain on the surface to be welded; the weld groove should be inspected according to the relevant requirements of the current national standard GB 50661 *Code for Welding of Steel Structures*.

6.3.6 The welding shall be carried out according to the welding parameters for the procedure qualification.

6.3.7 When the ambient temperature for welding is less than $0°C$ and not lower than $-10°C$, heating or protection measures shall be taken; as for the base material greater than or equal to 2 times of the steel plate thickness and not less than 100mm in directions of the welded joints and surfaces, it shall be welded after it is heated to the minimum preheat temperature that shall be not lower than $20°C$.

III Tack Welding

6.3.8 Thickness of the tack welded joint shall be not less than 3mm and should be not greater than 2/3 of the design weld thickness; the length should be not less than 40mm and 4 times of the thickness of the thinner part in the joint; the spacing should be 300mm~600mm.

6.3.9 Tack welds shall be subject to the same welding procedure and quality requirements as the final welds. Multiple-pass tack welds shall have cascaded ends. For the welded joint that the steel backup plate is adopted, the tack welding should be carried out in the bevel. During the tack welding, the preheat temperature should be 20℃~50℃ higher than the formal welding preheat temperature.

IV Run-on Tab, Run-off Tab and Backup Plate

6.3.10 When the run-on tab, run-off tab and backup plate are made of steel materials, the yield strength of the steel not greater than the nominal strength of the welded steel shall be selected and the weldability shall be close to each other.

6.3.11 Weld run-on tab and run-off tab shall be arranged at the ends of the welded joints. The lead-out length of the Shielded Metal Arc Welded and Gas Metal Arc Welded joints shall be greater than 25mm and that of the Submerged Arc Welded joint shall be greater than 80mm. After the welding is finished and completely cooled, the run-on tab and run-off tab may be removed by flame cutting, carbon arc air gouging or mechanical methods and they shall be ground flat and smooth; and it is strictly forbidden to knock them down with a hammer.

6.3.12 The steel backup plate shall be closely connected with the base material of the joint with the clearance not greater than 1.5mm and it shall be fully penetrated with the weld. When the manual Arc Welding and Gas Metal Arc Welding are adopted, the thickness of the steel backup plate shall be not less than 4mm; when the Submerged Arc Welding is adopted, that of the steel backup plate shall be not less than 6mm; when the Electroslag Welding is adopted, that of the steel backup plate shall be not less than 25mm.

V Preheat and Inter-pass Temperature Control

6.3.13 Heating methods such as electrical heating, flame heating and infrared heating should be adopted for the preheat and inter-pass temperature control and special thermodetector shall be adopted for the measurement. The preheating zone shall be located on both sides of the welding bevel and the width shall be more than 1.5 times of the plate thickness in the welded area of the welded part and shall be not less than 100mm. When it is a non-enclosed space member, the temperature measurement point should be at least 75mm away from the both sides of the welding bevel at the back of the heating surface for the welded part; when it is an enclosed space member, the temperature measurement point should be at least 100mm away from the welding bevel at the front surface.

6.3.14 The preheat and inter-pass temperature of welded joints shall meet the requirements of the current national standard GB 50661 *Code for Welding of Steel Structures*; when the preheat temperature selected for the procedure is less than those specified in the current national standard GB 50661 *Code for Welding of Steel Structures*, it shall be determined through the procedure qualification test.

VI Welding Deformation Control

6.3.15 The adopted welding procedure and sequence shall minimize the member deformation and shrinkage, the following welding sequences to control the deformation may be adopted:

1 Under permissible member placement conditions or easy turning, double-side symmetry welding should be adopted for the butt joint, Tee joint and cruciform joint; the member with symmetrical section should be welded in symmetry to its neutral axis; the connections with symmetrical connecting members should be welded in symmetry to its axis;

2 For the asymmetrical double groove weld, the weld on the deep bevel side should be partially welded at first and then that on the shallow bevel side shall be fully welded; finally the weld on the deep bevel side is completed. The symmetry welding cycles should be increased for the extra-thick plate;

3 Back-step welding, skip welding or multi-person symmetry welding methods should be adopted for the long weld.

6.3.16 When the member is welded, methods such as reserving welding shrinkage allowance or presetting the reverse deformation should be adopted to control the shrinkage and deformation. The shrinkage allowance and reverse deformation should be determined through calculation or test.

6.3.17 In assemblies, joints expected to have significant shrinkage should usually be welded before joints expected to have lesser shrinkage. They should also be welded with as little restraint as possible.

VII Post-weld Stress-relief Treatment

6.3.18 When the post-weld stress relief is specified in the design document or contract document, for the butt joints under tensile stress in the structure subject to the fatigue assessment or connections or members with dense weld, electric heater for local annealing and heating furnace for integral annealing should be adopted for the post-weld stress-relief treatment; where the purpose is to stabilize the structure dimension, vibration method may be adopted for stress relief.

6.3.19 The post-weld heat treatment shall meet the relevant requirements of the current professional standard JB/T 6046 *Welding Assembly for Carbon Steel and Low Alloy Steel-Post-welding Heat Treatment Method*. When the electric heater is adopted for the local stress-relief heat treatment, it shall meet the following requirements:

1 Heating equipment equipped with automatic temperature controller is used and its heating, temperature measurement and temperature control performances shall meet the service requirements;

2 The width of heating plate (strip) on each side surface of the member weld shall be at least 3 times of the steel plate thickness and shall be not less than 200mm;

3 Both sides of the member outside the heating plate (strip) should be covered with insulation materials.

6.3.20 When the stress of the intermediate welding layer is relieved by hammering method, a ball peen hammer or small-sized vibrating tool shall be adopted and the base materials at the edges of the root weld, cap weld or weld groove shall not be hammered.

6.3.21 When a vibration method is adopted for stress relief, the selection and technical requirements of the vibration aging process parameter shall meet the relevant requirements of the current professional standard JB/T 10375 *Recommended Practice for Vibration Stress Relief on Welding Structure*.

6.4 Welded Joints

I Full Penetration and Partial Penetration Welding

6.4.1 For the weld consists of butt and corner joints requiring for full penetration such as Tee joint, cruciform joint and corner joint, the dimension of the weld leg of the reinforced fillet weld shall be not less than $t/4$ [Figure 6.4.1(a)~Figure 6.4.1(c)]. The weld leg dimension of the joint between the web and upper flange of the crane beam or similar member designed with fatigue assessment requirements shall be $t/2$ and shall be not greater than 10mm [Figure 6.4.1(d)]. The permissible tolerance of the weld leg dimension is 0~4mm.

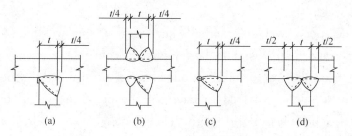

Figure 6.4.1 Dimension of Weld Leg

6.4.2 The weld reinforcement of the butt joint with full-penetration groove weld shall meet those specified in Table 6.4.2.

Table 6.4.2 Weld Reinforcement of Butt Joint(mm)

Designed weld grade	Weld width	Weld reinforcement
Grades I and II weld	<20	0~3
	≥20	0~4
Grade III weld	<20	0~3.5
	≥20	0~5

6.4.3 Bevel depth with different thicknesses may be adopted for the full-penetration double groove weld; the shallower bevel depth shall be not less than 1/4 the thickness of the joint.

6.4.4 Partial-penetration welding shall guarantee effective weld thickness required in the design document. As for the composite weld consisting of the partial-penetration groove weld and fillet weld in Tee joint and corner joint, the weld leg dimension of the reinforced fillet weld shall be 1/4 of the thickness of the thinnest plate in the joint and shall be not exceed 10mm.

II Fillet Welding

6.4.5 Components connected by the fillet weld shall be in close contact. The root space should be no greater than 2mm; when the root space of the joint is exceed 2mm, the weld leg dimension of the fillet weld shall be increased with the root space and the maximum shall be no greater than 5mm.

6.4.6 When ends of fillet weld are on the member, continuous boxing should be adopted for the corners. The distance from the arcing point/arc quenching point to the weld end shall be greater than 10.0mm; when continuous weld of the run-on and run-off tabs is not arranged at the ends of the fillet weld, the distance from the arcing/arc quenching point (Figure 6.4.6) to the weld end shall be greater than 10.0mm; the crater shall be filled up.

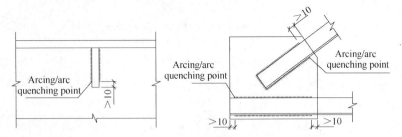

Figure 6.4.6 Position of the Arcing/Arc Quenching Point

6.4.7 The minimum length of each segment of the intermittent fillet weld shall be no less than 40mm; the maximum spacing between weld segments shall be no greater than 24 times of the thickness of the thinner welded part and shall be not greater than 300mm.

III Plug Welding and Slot Welding

6.4.8 Welding methods such as manual Shield Metal Arc Welding, Gas Metal Arc Welding and Self-Shielded Arc Welding may be adopted for the plug welding and the slot welding. For welds to be made in the flat position, fusing and depositing successive layers to fill the hole to the required depth. The slag shall be cooled and completely removed before restarting the weld. For welds to be made in the vertical position and overhead position, the slag should be allowed to cool and should be completely removed after depositing each successive bead until the hole is filled to the required depth.

6.4.9 The assembling clearance for the contact surface of two steel plates in plug welding and slot welding shall be no greater than 1.5mm. Infill plate is strictly forbidden to be used in the plug welding and slot welding.

IV Electroslag Welding

6.4.10 Special welding equipment shall be adopted for the Electroslag Welding; smelting nozzle and non-melting nozzle may be adopted for the welding. The steel backing and water cooling copper backing may be used for the Electroslag Welding backing.

6.4.11 Symmetrical welding should be adopted for the Electroslag Welding of Tee joint for inner diaphragms and panels of box member.

6.4.12 Continuous welding should be adopted for tack welding of the backup plate and base material for the Electroslag Welding.

V Stud Welding

6.4.13 Studs shall be welded with particular welding equipment. Welding procedure qualification test shall be carried out and the welding parameters shall be determined for the first stud welding.

6.4.14 Before each shift of welding, testing shall be performed for at least 3 studs and the formal welding may be carried out only after they are qualified in the inspection.

6.4.15 When the particular equipment is not avaliable for welding due to the limitation, Shielded Metal Arc Welding and Gas Metal Arc Welding may be adopted for the stud; the welding shall be carried out according to corresponding welding parameters and the weld dimension shall be determined through calculation.

6.5 Welding Quality Inspection

6.5.1 Dimension tolerance, appearance quality and internal quality of the weld shall be inspected according to the current national standards GB 50205 *Code for Acceptance of Construction Quality of Steel Structures* and GB 50661 *Code for Welding of Steel Structures*.

6.5.2 After the studs are welded, bend test shall be carried out for random inspection; after they are bended by 30°, the weld and heat-affected zone shall be free from any visiable crack.

6.6 Repair of Welding Defects

6.6.1 When they exceed corresponding quality acceptance criteria, defects of the weld metal or base material may be thoroughly removed by wheel grinding, carbon arc air gouging, gouging or mechanical methods. The surface in the repair area shall be cleaned thoroughly before welding.

6.6.2 Repair of weld defects shall meet the following requirements:

1 In case of oversize weld collar, excessive convexity, or excessive reinforcement, excessive weld metal shall be removed by grinding or carbon arc air gouging;

2 In case of excessive concavity of weld or crater, undersize welds, undercutting, additional weld should be conducted;

3 In the case of incomplete fusion, excessive weld porosity, or slag inclusions, unacceptable portion shall be removed and rewelded;

4 The extent and depth of the crack shall be ascertained by use of magnetic-particle inspection, penetrant inspection or other equally non-destructive methods; the crack and sound metal beyond each end of the crack shall be removed by grinding or carbon air gouging; the repair welding shall be carried out again after it is determined that the crack has been completely removed by penetrant inspection or magnetic-particle inspection. For the cracks in the weld with larger restraint, i crack-arrest holes should be drilled on both ends of the crack before the crack is removed by the carbon arc air gouging. For the repair of the welding crack, the welding engineer shall be notified to investigate and analyze the crack reason and they shall do that according to the process requirements after developing special repair process scheme;

5 Preheat temperature for the repair of weld defects shall be 30℃~50℃ higher than that for normal welding under the same conditions; low-hydrogen welding method and materials shall be adopted for the welding;

6 The repaired weld part shall be welded continuously; post-heating and insulation measures shall be applied for the intermittent welding;

7 Repair times of the defect on the same part of the weld should be no more than 2. When it is greater than 2, procedure qualification shall be carried out before the repair and the subsequent repair welding shall be carried out after the procedure qualification is deemed as acceptable. Magnetic-particle inspection or dye penetrant inspection shall be added for the welded area after the repair.

7 Connecting of Fasteners

7.1 General Requirements

7.1.1 This Chapter is applicable to connection of fasteners in the manufacturing and installation of steel structures such as ordinary bolt, tor-shear type high strength bolt, high strength heavy-hexagon head bolt, high strength bolts and rivets for bolted spherical node of steel space frame, tapping screw and nail.

7.1.2 Connection node and splicing joint of member fasteners shall be fastened after they are qualified in the inspection.

7.1.3 Anti-corrosion and fire-retardant coating shall be timely carried out according to the design document for the connection nodes and splicing joint of fasteners qualified in the acceptance. Materials such as anti-corrosion putty shall be adopted to seal the joint in exposure to corrosive medium.

7.1.4 The manufacturing and installation organizations of steel structures respectively shall carry out the mean slip coefficient test for the high strength bolt connected friction surface according to the relevant requirements of the current national standards GB 50205 *Code for Acceptance of Construction Quality of Steel Structures*. The results shall meet the design requirements. When the strength design is carried out for the high strength bolt connection according to the bearing-type connection or tension-type connection, the mean slip coefficient test for the friction surface may be omitted.

7.2 Preparation of Connecting Pieces and Friction Surface Treatment

7.2.1 The bolt holes of connecting pieces shall be prepared according to the relevant requirements of Chapter 8 in this Code; and the precision of the bolt holes, hole wall surface roughness as well as the allowable tolerance of hole diameter and hole spacing shall meet the relevant requirements of the current national standard GB 50205 *Code for Acceptance of Construction Quality of Steel Structures*.

7.2.2 When the spacing of the bolt holes is greater than the allowable tolerance specified in Article 7.2.1 of this Code, the electrodes matched with the base material may be adopted for repair welding; holing shall be carried out again after they are qualified in the non-destructive testing and the re-drilling quantity, subject to repair welding, in each group of holes shall be less than or equal to 20% of bolts in the group of holes.

7.2.3 The contact surface clearance between the friction surfaces of high strength bolts because of the plate thickness tolerance, manufacturing tolerance or installation tolerance shall be treated according to those specified in Table 7.2.3.

Table 7.2.3 Treatment for Clearance between Contact Surfaces

Items	Sketch	Treatment method
1		When $\Delta < 1.0$ mm, no treatment

Table 7.2.3(continued)

Items	Sketch	Treatment method
2	Grinding the inclined plane	When $\Delta = (1.0 \sim 3.0)$mm, one side of the thick plate shall be ground to 1 : 10 gentle slope to make allowance less than 1.0mm
3		When $\Delta > 3.0$mm, a backup plate of at most three layers is added with the thickness no less than 3mm; the material and friction surface treatment method shall be the same as that for the member

7.2.4 Treatment process for the friction surface at the high strength bolt connection may be selected according to the requirements of the design mean slip coefficient; the mean slip coefficient shall meet the design requirements. When the manual grinding wheel is adopted for polishing, the polishing direction shall be vertical to the forced direction and the polishing range shall be not smaller than 4 times of the diameter of the bolt hole.

7.2.5 The high strength bolt connection friction surface after surface treatment shall meet the following requirements:

1 The connection friction surface shall be dry and clean, free from flash, burr, welding spatter, crater, iron scale and dirt etc.;

2 Protection measures shall be taken for the friction surface after treatment and the friction surface shall not be marked;

3 When anti-rust treatment method is adopted for the friction surface, before the installation, the floating rust on the friction surface shall be removed with a fine wire brush vertical to the forced direction of the member.

7.3 Connection of Ordinary Fasteners

7.3.1 Ordinary bolt may be fastened with an ordinary spanner so that the contact surface of the connected piece, bolt head and nut are closely contacted with the member surface. The ordinary bolt shall be fastened firstly from the middle and then to both sides in symmetry. In addition, secondary screwing should be adopted for large joint.

7.3.2 When the ordinary bolt is regarded as a permanent connecting bolt, the fastening shall meet the following requirements:

1 Plain washers shall be set on sides of the bolt head and nut. The washers set on the side of the bolt head shall be no more than 2 while that set on the side of the nut shall be no more than 1;

2 For the bolted connection bearing dynamic load or that on important positions, when anti-looseness requirements are made in the design, nut or spring washer with anti-looseness device shall be adopted and the spring washer shall be set on the side of the nut;

3 For the beveled bolted connections such as I-steel and U-steel, bevel washer should be adopted;

4 The number of the bolts for the same connection joint shall be no less than 2;

5 Quantity of the exposed screw threads after the fastening of the bolt shall be no less than 2 and the fastening quality may be inspected by knocking with a hammer.

7.3.3 Specification and dimension of the rivet, tapping screw and nail adopted for connecting steel sheet shall be matched with the connected steel plate; the spacing and edge distance shall meet the requirements of the design document. Heads of the steel rivet and tapping screws shall be leaned against the thinner plate piece. The tapping screw, steel rivet and nail shall be closely fastened with the connecting steel plate with tidy appearance arrangement.

7.3.4 The prefabricated hole diameter (d_0) on the connecting plate for the tapping screws (non-taping and non-drilling screws) may be calculated according to the following formula:

$$d_0 = 0.7d + 0.2t_1 \qquad (7.3.4\text{-}1)$$
$$d_0 \leqslant 0.9d \qquad (7.3.4\text{-}2)$$

Where d——the nominal diameter of the tapping screw, mm;
　　　t_1——the total thickness of connecting plate, mm.

7.3.5 the penetration depth of nail connection shall be no less than 10.0mm.

7.4 Connection of High Strength Bolts

7.4.1 The set of high strength heavy hexagon-head bolt shall be composed of one bolt, one nut and two washers while the set of tor-shear type high strength bolt shall be composed of one bolt, one nut and one washer. The application combination shall meet those specified in Table 7.4.1.

Table 7.4.1　Application Combination of Sets of High Strength Bolt

Bolt	Nut	Washer
10.9S	10H	(35~45)HRC
8.8S	8H	(35~45)HRC

7.4.2 The length of the high strength bolt shall be calculated according to 2~3 exposed screw threads after the set of bolt is finally screwed; and it may be calculated according to the following formula. The length after rounding shall be taken for the nominal length of the selected high strength bolt and shall be rounded off at 5mm rounding interval according to the calculated bolt length l.

$$l = l' + \Delta l \qquad (7.4.2\text{-}1)$$
$$\Delta l = m + ns + 3p \qquad (7.4.2\text{-}2)$$

Where l'——the total thickness of the connecting plate layer;
　　　Δl——the additional length; or it is selected according to Table 7.4.2;
　　　m——the normal thickness of high strength nut;
　　　n——the quantity of washers; for the tor-shear type high strength bolt, 1 is taken; for the high strength heavy-hexagon head bolt, 2 is taken;
　　　s——the nominal thickness of the high strength washer; when large round hole or slotted hole is adopted, the nominal thickness of the high strength washer is taken according to the actual thickness;
　　　p——the thread pitch.

Table 7.4.2　Additional Length (Δl) of High Strength Bolt(mm)

Categories	Specifications						
	M12	M16	M20	M22	M24	M27	M30
High strength heavy-hexagon head bolt	23	30	35.5	39.5	43	46	50.5
Tor-shear type high strength bolt	—	26	31.5	34.5	38	41	45.5

Note: the additional length (Δl) in this table is determined by calculating the nominal thickness of the standard round hole washer.

7.4.3 When the high strength bolt is installed, the mounting bolt and drift pin shall be firstly used. The quantity of the mounting bolts and drift pins fitted at each node shall be determined according to the bearing load calculation in the assembly process, and shall meet the following requirements:

 1 It shall be no less than 1/3 of the total mounting holes;

 2 Quantity of the mounting bolts shall be no less than 2;

 3 Quantity of the fitted drift pins should be no more than 30% of that of the mounting bolts;

 4 The high strength bolt shall not be doubled as the mounting bolt.

7.4.4 The high strength bolt shall be tightened after the installation accuracy of the member is adjusted. The installation of the high strength bolt shall meet the following requirements:

 1 During the installation of the tor-shear type high strength bolt, the side of the nut with truncated cone shall face towards the side of the washer with chamfering;

 2 During the installation of the high strength heavy-hexagon head bolt, the side of the lower bolt head washer with chamfering shall face towards bolt head and the side of the nut with truncated cone shall face toward the side of the washer with chamfering.

7.4.5 During the site installation, the high strength bolt shall be able to be freely fitted in the bolt holes, but not in a forced way. When the bolt fails to be fitted, a reamer or file may be adopted to trim the bolt holes, but gas cutting shall not be adopted for reaming hole. The quantity of the reaming holes shall be approved by the design organization and the hole diameter after the trimming or reaming shall be no larger than 1.2 times of the diameter of the bolt.

7.4.6 The set of high strength heavy hexagon head bolt may be screwed by torque/rotation angle method and the screwing shall meet the following requirements:

 1 The torque spanner for construction shall be calibrated before the use; the relative torque error shall be no more than $\pm 5\%$ and that of the torque spanner for calibration shall be no more than $\pm 3\%$;

 2 During the screwing, a torque shall be applied to the nut;

 3 The screwing shall be divided into the primary screwing and final screwing. Secondary screwing shall be added for large nodes during the primary screwing and final screwing. 50% of the final screwing torque for construction may be taken for the primary screwing. The secondary screwing torque shall be equal to the primary screwing torque. The final screwing torque shall be calculated according to the following formula:

$$T_c = kP_c d \quad (7.4.6)$$

Where T_c——the final screwing torque for construction, N·m;

    k——the mean torque coefficient of the set of high strength bolt, 0.110~0.150 is taken;

    P_c——Pretension force for construction of the high strength heavy hexagon head bolt; it may be selected according to those specified in Table 7.4.6-1, kN;

    d——the normal diameter of the high strength bolt, mm;

Table 7.4.6-1 Pretension force for Construction of High Strength Large Hexagon Head Bolt (kN)

Bolt grade	Nominal diameter (mm)						
	M12	M16	M20	M22	M24	M27	M30
8.8S	50	90	140	165	195	255	310
10.9S	60	110	170	210	250	320	390

4 When the construction is carried out by rotation angle method, the final screwing angle of the set of bolt after primary screwing (secondary screwing) shall meet those specified in Table 7.4.6-2;

Table 7.4.6-2 Final Screwing Angle of the Set of Bolt after Primary Screwing (Secondary Screwing)

Bolt length, l	Rotation angle of nut	Connection state
$l \leqslant 4d$	1/3 cycle (120°)	The connection type is one layer of core plate with two additional layers of cover plates.
$4d < l \leqslant 8d$ or 200mm or below	1/2 cycle (180°)	
$8d < l \leqslant 12d$ or 200mm above	2/3 cycle (240°)	

Notes: 1 d is the nominal diameter of the bolt;

2 The rotation angle of the nut is the relative rotation angle between the nut and the bolt bar;

3 When the bolt length (l) is greater than 12 times of the nominal diameter (d) of the bolt, the final screwing angle of the nut shall be determined through a test.

5 The nut shall be marked in color after the primary screwing or secondary screwing.

7.4.7 Special electric wrench shall be adopted to screw the set of tor-shear type high strength bolt and the screwing shall meet the following requirements:

1 The screwing is divided into the primary screwing and final screwing; the secondary screwing should be increased for the large nodes between the primary screwing and final screwing;

2 50% the calculated value of T_c in Formula (7.4.6) of this Code shall be taken for the primary screwing torque; thereinto, 0.13 shall be taken for k and it may also be selected according to those specified in Table 7.4.7; the secondary screwing torque shall equal to the primary screwing torque;

Table 7.4.7 Primary Screwing (Secondary Screwing) Torque of the Tor-shear Type High Strength Bolt(N·m)

Nominal diameter (mm)	M16	M20	M22	M24	M27	M30
Primary screwing (secondary screwing) torque	115	220	300	390	560	760

3 The final screwing shall be subject to the wobblers with the bolt tail twisted off. Final screwing for few bolts that the final screwing can't be carried out with a special spanner may be carried out by methods specified in Article 7.4.6 of this Code. In addition, 0.13 shall be taken for the torque coefficient (k).

4 The nut shall be marked in color after the primary screwing or secondary screwing.

7.4.8 Reasonable screwing sequence shall be adopted for the primary screwing, secondary screwing and final screwing of the high strength bolt connection groups.

7.4.9 When it is not specified in the design document, the connection node for combined use of the high strength bolt and welding should be subject to the construction sequence of first bolt fastening and then welding.

7.4.10 The primary screwing, secondary screwing and final screwing of the set of high strength bolt should be completed within 24h.

7.4.11 When the high strength heavy-hexagon head bolt is fastened by the torque method, the following quality inspection shall be carried out:

1 Final screwing color mark shall be inspected and a 0.3kg hammer shall be used to knock the nut for inspecting the high strength bolt one by one;

2 The final screwing torque shall be randomly inspected according to 10% of the quantity of nodes (at least 10 nodes); for each node, it shall be randomly inspected according to 10% of the quantity of bolts (at least 2 bolts);

3 During the inspection, a straight line shall be drawn firstly on the end face of the screw rod and the nut; then the nut is unscrewed by about 60° and tightened up again with a torque spanner to make the two lines superposed; the measured torque on this occasion shall be $0.9T_{ch} \sim 1.1T_{ch}$. The T_{ch} may be calculated according to the following formula:

$$T_{ch} = kPd \qquad (7.4.11)$$

Where T_{ch}——the inspected torque, N·m;
  P——the design pretension force of the high strength bolt, kN;
  k——the torque coefficient.

4 In case of any nonconformity with the requirements, it shall be doubled for inspection. If it still fails to be qualified, the high strength bolts at the whole nodes shall be screwed again;

5 The torque inspection should be completed after 1h and before 24h final screwing of the bolt; and the relative error of the torque spanner for inspection shall be no more than ±3%.

7.4.12 When the high strength large hexagon head bolt is fastened by the rotation angle method, the following quality inspection shall be carried out:

1 Final screwing color mark shall be inspected and an about 0.3kg hammer shall be used to knock the nut for inspecting the high strength bolt one by one;

2 The final screwing rotation angle shall be randomly inspected according to 10% of the quantity of nodes (at least 10 nodes); for each node, it shall be randomly inspected according to 10% of the quantity of bolts (at least 2 bolts);

3 A line shall be drawn at the relative position of the screw rod end face and the nut. Then, all of the nuts are loosened and the bolt is tightened up again according to the specified primary screwing torque and final screwing angle; the angle between the end line and the original end line is measured and it shall meet those specified in Table 7.4.6-2; if the error is within ±30°, it shall be deemed as acceptable;

4 In case of any nonconformity with the requirements, it shall be doubled for inspection. If it still fails to be qualified, the high strength bolts at the whole nodes shall be screwed again;

5 Rotation angle inspection should be completed after 1h and before 24h final screwing of the bolt.

7.4.13 For the final screwing inspection of the tor-shear type high strength bolt, it shall be deemed as acceptable only after the tail wobbler is twisted off by visual inspection. For the tor-shear type high strength bolt that can't be tightened up with a special spanner, quality inspection shall be carried out according to the requirements of Article 7.4.11 in this Code.

7.4.14 After the assembling of the space frame for the bolted spherical node is completed, the high strength bolt and the spherical node shall be firmly fastened; the thread length of the bolt screwed in the bolted spheres shall be no less than 1.1 times of the bolt diameter and it shall be free from clearance and looseness at the fastening place.

8 Processing of Steel Parts and Components

8.1 General Requirements

8.1.1 This Chapter is applicable to the processing of parts and components in the fabrication of the steel structures.

8.1.2 Before the processing of parts and components, the design document and detail drawing for construction shall get acquainted and process for each procedure shall be prepared well; the processing document shall be prepared in combination with the actual processing conditions.

8.2 Setting out and Marking-off

8.2.1 The setting out and marking-off shall be carried out according to the detail drawing for construction and process document; and an allowance shall be reserved as required.

8.2.2 The allowable tolerance for the setting out and sample plate (sample rod) shall meet those specified in Table 8.2.2.

Table 8.2.2 Allowable Tolerance for Setting Out and Sample Plate (Sample Rod)

Items	Allowable Tolerance
Distance between parallel lines and segment dimension	±0.5mm
Length of sample plate	±0.5mm
Width of sample plate	±0.5mm
Difference between diagonal lines of sample plate	1.0mm
Length of sample rod	±1.0mm
Angle of sample plate	±20′

8.2.3 Allowable Tolerance for the marking-off shall meet those specified in Table 8.2.3.

Table 8.2.3 Allowable Tolerance for Marking-off(mm)

Items	Allowable Tolerance
Overall dimension of part	±1.0
Hole spacing	±0.5

8.2.4 For the major parts, the marking-off shall be carried out based on the load condition and processing condition of the members, along the the direction specified in the process.

8.2.5 After the marking-off, the parts and components shall be marked according to the detail drawing for construction and process requirements.

8.3 Cutting

8.3.1 Gas cutting, mechanical cutting and plasma cutting methods may be adopted for the cutting of the steel materials; the selected cutting method shall meet the requirements of the process document and the flash and burr after cutting shall be cleaned up.

8.3.2 The cut surface of the steel materials shall be free from defects such as crack, slag inclusion and lamination as well as more than 1mm edge defect.

8.3.3 Before gas cutting, the surface of the steel materials in the cutting area shall be cleaned up. During the cutting, appropriate process parameter shall be selected according to factors such as the equipment type, thickness of the steel materials and cutting gas.

8.3.4 Allowable tolerance for the gas cutting shall meet those specified in Table 8.3.4.

Table 8.3.4 Allowable Tolerance for Gas Cutting(mm)

Items	Allowable tolerance
Width and length of parts	±3.0
Planeness of the cut surface	$0.05t$, and $\leqslant 2.0$
Cut depth	0.3
Local notch depth	1.0

Note: t is the thickness of the cut surface.

8.3.5 The thickness of the parts under mechanical shearing should be no more than 12.0mm; the cut surface shall be flat and smooth. The carbon structural steels shall not be cut and punched when the ambient temperature is less than −20℃, neither shall the low alloy structural steel when the ambient temperature is less than and −15℃.

8.3.6 The allowable tolerance for mechanical shearing shall meet those specified in Table 8.3.6.

Table 8.3.6 Allowable Tolerance for Mechanical Shearing(mm)

Items	Allowable tolerance (mm)
Width and length of parts	±3.0
Edge defect	1.0
Perpendicularity of steel section end	2.0

8.3.7 For the steel pipe member bars in steel space frame (truss), pipe lathe or CNC intersecting line cutting machine should be used for material preparation. Machining allowance and welding shrinkage shall be included in advance in the material preparation. The welding shrinkage may be determined by the process test. The allowable tolerance for the processing of the steel pipe member bar shall meet those specified in Table 8.3.7.

Table 8.3.7 Allowable Tolerance for Processing of Steel pipe Member Bar(mm)

Items	Allowable Tolerance
Length	±1.0
Perpendicularity of the end face to the pipe axis	$0.005r$
Curve at the pipe orifice	1.0

Note: r is radius of the tube.

8.4 Rectification and Formation

8.4.1 Mechanical and/or heating methods may be adopted for the rectification.

8.4.2 When the ambient temperature is less than −16℃ for the carbon structural steels and is less than −12℃ for the low alloy structural steels, cold rectification and cold bending shall not be carried out. During the heating rectification of the carbon structural steels and low alloy structural steels, the heating temperature shall be 700℃∼800℃; the maximum temperature is strictly forbidden to exceed 900℃ and the minimum temperature shall be at least 600℃.

8.4.3 When the thermal formation is adopted for parts, different heating temperatures may be selected according to the carbon content of the materials. The heating temperature shall be controlled within 900℃~1000℃ or 1100℃~1300℃. The processing shall be finished before the temperatures of the carbon structural steels and low alloy structural steels fall to 700℃ and 800℃, respectively; low alloy structural steels shall cool down naturally.

8.4.4 The temperature of the thermal formation shall be uniform and hot processing shall not be carried out repeatedly for the same member; when the temperature falls to 200℃ ~ 400℃, hammering, bending and formation are strictly forbidden.

8.4.5 For the factory cold-formed steel pipe, rolling or pressing process may be adopted.

8.4.6 Surfaces of the steel materials after the rectification shall be free from obvious dent or damage; the scratch depth shall be no more than 0.5mm and 1/2 negative allowable tolerance for the thickness of steel materials.

8.4.7 Minimum curvature radius and maximum bending rise for the cold rectification and cold bending of profile steel shall meet those specified in Table 8.4.7.

Table 8.4.7 Minimum Curvature Radius and Maximum Bending Rise for Cold Rectification and Cold Bending (mm)

Type of the steel materials	Sketch	Corresponding axis	Rectification r	Rectification f	Bending r	Bending f
Flat steel		x-x	$50t$	$\dfrac{l^2}{400t}$	$25t$	$\dfrac{l^2}{200t}$
Flat steel		y-y (Only applicable to flat steel axis)	$100b$	$\dfrac{l^2}{800b}$	$50b$	$\dfrac{l^2}{400b}$
Angle steel		x-x	$90b$	$\dfrac{l^2}{720b}$	$45b$	$\dfrac{l^2}{360b}$
U-steel		x-x	$50h$	$\dfrac{l^2}{400h}$	$25h$	$\dfrac{l^2}{200h}$
U-steel		y-y	$90b$	$\dfrac{l^2}{720b}$	$45b$	$\dfrac{l^2}{360b}$
I-steel		x-x	$50h$	$\dfrac{l^2}{400h}$	$25h$	$\dfrac{l^2}{200h}$
I-steel		y-y	$50b$	$\dfrac{l^2}{400b}$	$25b$	$\dfrac{l^2}{200b}$

Note: r is the curvature radius, f is the bending rise, l is the bending chord length, t is the plate thickness, b is the width and h is the height.

8.4.8 Allowable tolerance of the steel materials after the rectification shall meet those specified in Table 8.4.8.

Table 8.4.8 Allowable tolerance of Steel Materials after the Rectification(mm)

Items		Allowable tolerance	Sketch
Local planeness of steel plates	$t \leqslant 14$	1.5	
	$t > 14$	1.0	
Bending rise of profile steel		$l/1000$, and $\leqslant 5.0$	
Perpendicularity of angle steel leg		$b/100$, and the angle of the double-leg bolted angle steel shall be no larger than 90°.	
Perpendicularity of the U-steel flange to the web plate		$b/80$	
Perpendicularity of the I-steel and H steel flanges to the web plate		$b/100$, and $\leqslant 2.0$	

8.4.9 Allowable tolerance for the bending and formation of steel pipe shall meet those specified in Table 8.4.9.

Table 8.4.9 Allowable Tolerance for Bending and Formation of Steel Pipe(mm)

Items	Allowable tolerance
Diameter	$\pm d/200$, and $\leqslant \pm 5.0$
Member length	± 3.0
Roundness of pipe orifice	$d/200$, and $\leqslant 5.0$
Roundness at the middle of the pipe	$d/100$, and $\leqslant 8.0$
Bending rise	$l/1500$, and $\leqslant 5.0$

Note: d is the diameter of the steel pipe.

8.5 Trimming of Edges

8.5.1 Gas cutting and machining method may be adopted for the trimming of edges and precision cutting shall be adopted for edges with special requirements.

8.5.2 When the trimming of edges needs to be carried out for the gas cutting or mechanical shearing parts, the planing thickness shall be no less than 2.0mm.

8.5.3 Allowable tolerance for the trimming of edges shall meet those specified in Table 8.5.3.

Table 8.5.3 Allowable Tolerance for Trimming of Edges

Items	Allowable tolerance
Width and length of parts	±1.0mm
Straightness of processed edge	$l/3000$, and ≤ 2.0mm
Included angle between adjacent sides	±6′
Perpendicularity of processed surface	$0.025t$, and ≤ 0.5mm
Surface roughness of the processed surface	Ra≤50μm

8.5.4 Gas cutting, skimming and processing with an edge planer may be adopted for the weld groove and the allowable tolerance of the weld groove shall meet those specified in Table 8.5.4.

Table 8.5.4 Allowable Tolerance of Weld Groove

Items	Allowable tolerance
Groove angle	±5°
Root face	±1.0mm

8.5.5 When a miller is adopted for the milling of the part edges, the allowable tolerance after the processing shall meet those specified in Table 8.5.5.

Table 8.5.5 Allowable Tolerance of Spare Parts after Milling(mm)

Items	Allowable tolerance
Length and width of parts when both ends are milled	±1.0
Planeness of milling surface	0.3
Perpendicularity of milling surface	$l/1500$

8.6 Holing

8.6.1 Methods such as drilling, punching, milling, reaming, boring and spot facing may be adopted for holing; gas cutting may also be adopted for holing with relatively large diameter or long hole.

8.6.2 When a driller is used for drilling hole in sandwich panel, effective measures to avoid play shall be taken.

8.6.3 After holing by mechanical or gas cutting methods, foreign bodies such as the burr and chip around the hole shall be removed; the hole wall shall be round and smooth as well as free from crack and more than 1.0mm edge defect.

8.7 Processing of Bolted Spheres and Welded Hollow Spheres

8.7.1 Hot forging should be adopted for the formation of bolted spheres; heating temperature shall be 1150℃~1250℃. The final forging temperature shall be no less than 800℃. The bolted spheres after formation shall be free from crack, wrinkle and overburning.

8.7.2 Allowable tolerance for processing of bolted spheres shall meet those specified in Table 8.7.2.

Table 8.7.2 Allowable Tolerance for Processing of Bolted Spheres (mm)

Items		Allowable tolerance
Diameter	$d \leqslant 120$	+2.0 / −1.0
	$d > 120$	+3.0 / −1.5
Roundness	$d \leqslant 120$	1.5
	$120 < d \leqslant 250$	2.5
	$d > 250$	3.0
Parallism of two milling surfaces on the same axis	$d \leqslant 120$	0.2
	$d > 120$	0.3
Distance from milling surface to the sphere center		±0.2
Included angle between two adjacent bolt hole center lines		±30′
Perpendicularity of two milling surfaces to the bolt hole axis		$0.005r$

Note: r is the radius of the bolted sphere and d is its diameter.

8.7.3 The welded hollow sphere shall be hot pressed steel plate into semi-round sphere. The heating temperature shall be 1000℃~1100℃ and it shall be welded round sphere after the groove machining. The surface of the finished sphere after welding shall be flat and smooth, free from local projection or wrinkle.

8.7.4 Allowable tolerance for processing of the welded hollow sphere shall meet those specified in Table 8.7.4.

Table 8.7.4 Allowable Tolerance for Processing of Welded Hollow Spheres (mm)

Items		Allowable tolerance
Diameter	$d \leqslant 300$	±1.5
	$300 < d \leqslant 500$	±2.5
	$500 < d \leqslant 800$	±3.5
	$d > 800$	±4
Roundness	$d \leqslant 300$	±1.5
	$300 < d \leqslant 500$	±2.5
	$500 < d \leqslant 800$	±3.5
	$d > 800$	±4
Wall thickness reduction	$t \leqslant 10$	$\leqslant 0.18t$, and $\leqslant 1.5$
	$10 < t \leqslant 16$	$\leqslant 0.15t$, and $\leqslant 2.0$
	$16 < t \leqslant 22$	$\leqslant 0.12t$, and $\leqslant 2.5$
	$22 < t \leqslant 45$	$\leqslant 0.11t$, and $\leqslant 3.5$
	$t > 45$	$\leqslant 0.08t$, and $\leqslant 4.0$
Butt misalignment	$t \leqslant 20$	$\leqslant 0.10t$, and $\leqslant 1.0$
	$20 < t \leqslant 40$	2.0
	$t > 40$	3.0
Weld reinforcement		0~1.5

Note: d is the outside diameter of the welded hollow sphere and t is its wall thickness.

8.8 Processing of Cast Steel Node

8.8.1 Casting process and processing quality of the cast steel node shall meet the requirements of the design document and current relevant standards of the nation.

8.8.2 Processing of the cast steel node should cover procedures such as process design, modeling, pouring, cleaning, heat treatment, polishing (repair), machining and inspection of finished product.

8.8.3 The complex cast steel node joint should be arranged with transition section.

8.9 Processing of Rope Nodes

8.9.1 Rope nodes may be processed into a blank by methods such as casting, forging and welding, subject to the machining such as turning, milling, planing, drilling and boring.

8.9.2 Ordinary thread of rope node shall meet the requirements of the current national standards GB/T 196 *General Purpose Metric Screw Threads-Basic Dimensions* and GB/T 197 *General Purpose Metric Screw Threads-Tolerances* with regard to 7H/6g. The trapezoidal thread shall meet the requirements of the current national standard GB/T 5796 *Trapezoidal Screw Threads* with regard to the 8H/7e.

9 Assembling and Processing of Members

9.1 General Requirements

9.1.1 This Chapter is applicable to assembling and processing of members in the fabrication and installation process of the steel structures.

9.1.2 Before the member assembling, the assembler shall get familiar with the detail drawing for construction, assembling process and relevant requirements of technical documents. The materials, specification, appearance, dimension and quantity of parts for assembling inspection shall meet the design requirements.

9.1.3 The iron rust, burr and dirt on the contact surface at the assembly welding place and within 30mm~50mm of edges shall be cleaned up before assembling.

9.1.4 Assembling of panels and profiles shall be carried out before the member assembling; the member assembling shall be carried out after the component is subject to assembling, welding and calibration as well as qualified in the inspection.

9.1.5 Reasonable member assembling sequence shall be determined according to the design requirements, member type, connecting mode, welding method and welding sequence.

9.1.6 Concealed position of the member shall be closed after the welding and coating are qualified in the inspection; and coating may be omitted for the internal surface of the fully-closed member.

9.1.7 Member shall be welded after the assembling is completed and qualified in the inspection.

9.1.8 End face processing shall be carried out for the welded member according to the requirements of the design and process documents.

9.1.9 The dimension tolerance of member assembling shall meet the relevant requirements of the design document and the current national standard GB 50205 *Code for Acceptance of Construction Quality of Steel Structures*.

9.2 Assembling of Components

9.2.1 The spacing of the assembling joints of steel flange plate and the web plate in welded H-shape steel section shall be no less than 200mm. The assembling length of the flange plate shall be no less than 600mm; and the assembling width and length of the web plate shall be no less than 300mm and 600mm, respectively.

9.2.2 Fox box section steel, the assembling length of the side plate shall be no less than 600mm; the spacing between assembling joints of two adjacent side plates should be no less than 200mm; the side plate should not be assembled in the width direction; when the width is greater than 2400mm and the assembling is necessary, the minimum assembling width should be no less than 1/4 of the plate width.

9.2.3 When no special requirement is made in the design, straight-opening full-penetration welded assembling may be adopted for the rolled profile steel for secondary member and the assembling length shall be no less than 600mm.

9.2.4 One joint should be provided between nodes when the steel pipe is extended and the minimum extension length shall meet the following requirements:

    1 When the diameter (d) of the steel pipe is no greater than 500mm, it shall be no less than 500mm;

    2 When the diameter (d) of the steel pipe is greater than 500mm and less than or equal to 1000mm, it shall be no less than d;

    3 When the diameter (d) of the steel pipe is greater than 1000mm, it shall be no less than 1000mm;

    4 When the steel pipe is formed by rolling, several joints may be provided, but the minimum extension length shall meet requirements of Items 1~3 in this article.

9.2.5 When the steel pipe is extended, the longitudinal weld of adjacent pipe joints or sections shall be staggered and the minimum staggered distance (along the direction of the arc length) shall be no less than 5 times of the steel pipe wall thickness and shall be no less than 200mm.

9.2.6 The butt weld of components shall meet the requirements of the design document; when it is not specified in the design, full penetration butt weld shall be adopted.

9.3 Assembling of Members

9.3.1 The member should be assembled on the assembly platform, support or special equipment. The assembly platform and support shall not only have adequate strength and stiffness, but also be convenient for assembling/disassembling and positioning. On the assembly platform or support, reference lines such as the center line of the member, position line of end face, contour line and elevation line should be drawn.

9.3.2 Sample plot, profiling, forming die and special equipment assembling methods may be adopted for the member assembly; and vertical and horizontal modes may be adopted for the assembling.

9.3.3 The assembling clearance of the member shall meet the requirements of the design and process documents. When it is not specified in the design and process documents, the assembling clearance should be no greater than 2.0 mm.

9.3.4 When the welded member is assembled, the welding shrinkage shall be preset and reasonable welding shrinkage shall be distributed for the components. The welding shrinkage should be determined through process test for the important or complex member.

9.3.5 When the arching is specified for the member in the design, the arching shall be carried out according to the specified arching value during the assembling and the allowable tolerance for the arching is 0~10% of the arching value and shall be no greater than 10mm. When it is not specified in the design, but specified in the construction process, the allowable tolerance for the arching shall be no greater than ±10% of the arching value and shall be no greater than ±10mm.

9.3.6 When the truss structure is assembled, the displacement of the intersection point between the centers line of the member bar shall be no greater than 3mm.

9.3.7 After the crane beam and crane truss are assembled and welded, they are not permissible for down-warping. The lower flange of the crane beam and tension surface of important loaded member shall not be welded with tooling fixture as well as temporary positioning plate and connecting plate.

9.3.8 Temporary tooling fixture, temporary positioning plate and temporary connecting plate are removed; they are strictly forbidden to be knocked down with a hammer, but shall be removed by gas cutting mode at a distance of 3mm~5mm away from the member surface; the residual crater shall be polished flat and smooth and it shall not damage the base material.

9.3.9 After the member ends are milled, over 75% area shall be in close contact with the contact surface; 0.3mm feeler gauge shall be adopted for the inspection. The penetrated area shall be less than 25% and the maximum clearance between edges shall be no greater than 0.8mm.

9.4 Milling of Member Ends

9.4.1 Milling of member ends shall be carried out after the member is assembled, welded and qualified in the inspection; and end milling machine may be used for the milling of member ends.

9.4.2 Milling of member ends shall meet the following requirements:

   1 Milling amount of ends shall be predetermined according to the process requirements and it should be no less than 5mm;

   2 Planeness and perpendicularity of the milling surface shall be controlled according to the relevant requirements of the design document and the current national standard GB 50205 *Code for Acceptance of Construction Quality of Steel Structures*.

9.5 Rectification of Members

9.5.1 The member shape should be rectified according to sequences from the overall to the local, from the primary to the secondary and from the lower to the upper.

9.5.2 Cold and hot rectification may be adopted for the rectification of member shape. When it is specified in the design, the rectification and temperature shall meet requirements of the design document. When it is not specified in the design, the rectification method and temperature shall meet requirements of Section 8.4 in this Code.

10 Test Assembling of Steel Structures

10.1 General Requirements

10.1.1 This Chapter is applicable to test assembling of members specified in the Contract or design document.

10.1.2 Before test assembling, individual member shall be qualified in the inspection; when quantity of the same type of members is relatively large, certain representative members may be selected for test assembling.

10.1.3 Integral test assembling or accumulated and continuous test assembling may be adopted for the member. When the latter is adopted, the members connected by two adjacent units shall participate in test assembling of the two units respectively.

10.1.4 Unless otherwise specified, acceptance shall be carried out for the test assembling for the member according to the design document and the current national standard GB 50205 *Code for Acceptance of Construction Quality of Steel Structures*. During the test assembling acceptance, it shall be kept clear from the sunlight.

10.2 Test Assembling of Steel Structures

10.2.1 The test assembling site shall be flat, smooth and solid; the temporary bearing support, stool or platform for the test assembling shall be exactly positioned through measurement and shall meet the requirements of the process document. Structural safety checking shall be carried out for the temporary structure for test assembling of heavy member.

10.2.2 Suitable geometric shape may be selected according to site conditions and hoisting equipment for the test assembling of test assembling units.

10.2.3 Test assembling shall be carried out for the member in the free state.

10.2.4 Test assembling of the member shall be positioned according to the control dimension in the design drawing. For test assembling members with pre-arching and welding shrinkage, the sizing shall be adjusted according to the pre-arching value or shrinkage.

10.2.5 Where necessary, drilling may be carried out for connecting pieces of nodes that are connected by bolts after the positioning of the test assembling.

10.2.6 When sandwich panel stacks are connected with high strength bolts or ordinary bolts, at least drift pins of 10% of the total bolt holes should be firstly used for positioning and then they are fastened with temporary bolts. Quantity of temporary bolts in one group of holes shall be no less than 20% that of the bolt holes and 2. During the test assembling, the panels shall be fitted. A hole tester shall be adopted to inspect the bolt holes and they shall meet the following requirements:

    1 When a hole tester whose nominal diameter is 1.0mm less than that of the hole for inspection, the through rate of each group of holes shall be no less than 85%;

    2 When a hole tester whose nominal diameter is 0.3mm greater than that of the bolt for inspection, the through rate shall be 100%.

10.2.7 After the test assembling is qualified in the inspection, center line and control reference line should be marked on the member; where necessary, a positioner may be arranged.

10.3 Test Assembling of Computer Assistance Simulation

10.3.1 Both the test assembling of steel structures and the test assembling of computer assistance simulation may be adopted for the member; the overall dimension of the simulation member or unit shall be the same as the geometric dimension of the physical object.

10.3.2 When the tolerance due to the test assembling of computer assistance simulation is greater than that specified in the current national standard GB 50205 *Code for Acceptance of Construction Quality of Steel Structures*, the test assembling of steel structures shall be carried out according to the requirements of Section 10.2 in this Code.

11 Installation of Steel Structures

11.1 General Requirements

11.1.1 This Chapter is applicable to installation of single-story steel structure, multi-story and tall steel structures, long-span spatial structures and high-rising steel structures.

11.1.2 The installation site of steel structures shall be arranged with special member stockyard and protection measures shall be taken to avoid deformation and surface contamination of member.

11.1.3 Before erection, the approaching members shall be checked according to the list of members and the product certificate also shall be checked; when the member subject to the test assembling in the factory is assembled on the site, it shall be carried out according to the test assembling record.

11.1.4 Before the member hoisting, foreign bodies such as oil stain, ice/snow, silt and dust on the surface shall be removed; and the axis and elevation shall be well marked.

11.1.5 The installation of steel structures shall be carried out in reasonable sequence according to the structure characteristics; and it shall form stable space rigid unit; where necessary, temporary structure or temporary measure shall be added.

11.1.6 During the calibration for the installation of steel structures, influences of factors such as temperature, sunlight and welding deformation on the structural deformation shall be analyzed. The construction organization and supervision organization should carry out the measurement and acceptance under the same weather conditions and duration.

11.1.7 During the hoisting of steel structures, special hoisting lug or hoisting hole should be arranged on the member. When it is not particularly specified in the design document, the hoisting lug and hoisting hole may be reserved on the member. When it is necessary to remove the hoisting lug, gas cutting or carbon arc air gouging may be adopted to cut it at the place of 3mm~5mm away from the base material, but hammering is strictly forbidden to be adopted for removal.

11.1.8 During the installation of steel structures, holing, assembling, welding and coating shall meet the relevant requirements of Chapters 6, 8, 9 and 13 in this Code.

11.1.9 For the member coating damaged in the process of transportation, storage and installation as well as the installation connecting parts, paint repair shall be carried out according to the relevant requirements of Chapter 13 in this Code.

11.2 Hoisting Equipment and Hoisting Mechanism

11.2.1 Approved products such as tower crane, crawler crane and auto-crane should be adopted for the installation of steel structures. When non-approved products are selected as the hoisting equipment, specific plan shall be prepared and be implemented after the review.

11.2.2 The hoisting equipment shall be determined comprehensively according to factors such as performance, structure characteristics, site environment and operation efficiency.

11.2.3 When the hoisting equipment is required to be attached to or borne by the structure, consent from the design organization shall be obtained and the structural safety shall be checked.

11.2.4 Hoisting of the steel structures must be carried out within the rated hoisting capacity of the hoisting equipment.

11.2.5 Lifting should not be adopted for the hoisting of steel structures. When the member weight exceeds the range of the rated hoisting capacity of single hoisting equipment, lifting mode may be adopted for the hoisting of the member. When lifting mode is adopted, it shall meet the following requirements:

    1 Reasonable load distribution shall be carried out for the hoisting equipment. The member weight shall be less than or equal to 75% of total rated hoisting capacity of the two hoisting equipment and the load capacity of single hoisting equipment shall be less than or equal to 80% of the rated hoisting capacity.

    2 Safety checking shall be carried out and corresponding safety measures shall be taken for the hoisting. In addition, an approved specific lifting plan shall be provided;

    3 During the hoisting, the two hoisting equipment shall maintain synchronous lifting and movement; and their hooks and pulley block shall basically maintain mutual verticality.

11.2.6 Hoisting mechanisms such as hoisting rope, belt, shackle and hook shall be qualified in the inspection and be used within the range of the rated permissible load.

11.3 Foundation, Bearing Surface and Embedment Parts

11.3.1 Before the installation of steel structures, positioning axis, foundation axis and elevation as well as anchor bolts position shall be inspected; and handover and acceptance shall be carried out. When the foundation engineering is handed over in batches, column foundations of at least one installation unit shall be subject to each handover and acceptance and shall meet the following requirements:

    1 Strength of foundation concrete shall meet the design requirements;

    2 Backfilling and compaction around the foundation shall be completed;

    3 Foundation axis mark and elevation reference point shall be accurate and complete.

11.3.2 When the top surface of the foundation is directly regarded as the bearing surface of the column and the embedded steel plate (support) on the top surface of the foundation regarded as the bearing surface of the column, the allowable tolerance of the bearing surface and anchor bolt (anchor) shall meet those specified in Table 11.3.2.

Table 11.3.2 Allowable Tolerance of Bearing Surface and Anchor Bolt (Anchor) (mm)

Items		Allowable tolerance
Bearing surface	Elevation	±3.0
	Levelness	1/1000
Anchor bolt (anchor)	Displacement of bolt center	5.0
	Exposed length of the bolt	+30.0 0
	Thread length	+30.0 0
Displacement of the reserved hole center		10.0

11.3.3 When the steel shim plate is adopted as the bearing of the steel column base, it shall meet the following requirements:

1 The area of the steel shim plate shall be determined through calculation according to the concrete compression strength, the bearing load of the column base plate and fastening tension of anchor bolts (anchors);

2 The shim plate shall be arranged near the stiffening plate for the column base plate of the anchor bolt (anchor) or below the column component; 1~2 groups of shim plates shall be arranged on the side of the anchor bolt (anchor) and quantity of the shim plates in each group shall be less than or equal to 5;

3 The contact between the shim plate and the foundation surface/column bottom surface shall be flat, smooth and tight; when pairs of inclined bearing shim plates are adopted, the overlapping length shall be greater than or equal to 2/3 of the length of the shim plate;

4 Before the secondary concrete pouring for the column bottom, the shim plates shall be welded and fixed.

11.3.4 Installation of anchors and embedment parts shall meet the following requirements:

1 Auxiliary fixing facilities such as positioning support and positioning plate should be adopted for the anchors;

2 After the anchors and embedment parts are installed in place, they shall be reliably fixed; when the embedding precision of the anchors is relatively high, reserved hole and secondary embedding processes may be adopted;

3 Protection measures to avoid damage, rust and contamination shall be taken for the anchors;

4 After the anchor bolts for the steel column are fastened, the protection measures to avoid nut looseness and rust shall be taken for the exposed part;

5 When a prestress needs to be applied to the anchors, post-tensioning method may be adopted and the tension force shall meet the requirements of the design document; in addition, grouting treatment shall be carried out after the completion of the tension.

11.4 Installation of Members

11.4.1 Installation of the steel column shall meet the following requirements:

1 During the installation of the column base, an importer or sheath should be used for the anchors;

2 After the first section of the steel column is installed, perpendicularity, elevation and axis position calibration shall be timely carried out. The perpendicularity of the steel column may be measured with a theodolite or plumb; after the calibration is qualified, the steel column shall be reliably fixed and secondary grouting shall be carried out for the column bottom. Before the grouting, foreign bodies between the column base plate and foundation surface shall be removed.

3 The positioning axis of the steel column above the first section shall be directly led from the ground control axis, but not from the axis of the column at the lower layer; when the perpendicularity of the steel column is calibrated, the welding shrinkage of the steel beam joint shall be determined and weld shrinkage deformation value shall be reserved;

4 The inclined steel column may be measured and calibrated by three-dimensional coordinate measurement method or the combination of the projection point and the elevation. After the calibration is qualified, it should be fixed with a rigid brace.

11.4.2 Installation of the steel beam shall meet the following requirements:

1 Two-point hoisting should be adopted for the steel beam; when the length of single steel beam is greater than 21m and two-point hoisting can't meet the member strength and deformation requirements, 3~4 hoisting points should be arranged for hoisting or a balance beam should be adopted for hoisting; positions of the hoisting points shall be determined through calculation;

2 One-machine-one (multiple)-hoisting may be adopted for the hoisting of the steel beam; after it is positioned, it shall be temporary fixed and connected immediately.

3 Elevation of the steel beam surface and the height difference on both ends may be measured with a water level and ruler; after the completion of the calibration, permanent connection shall be carried out.

11.4.3 Installation of the brace shall meet the following requirements:

1 The cross brace should be combined for hoisting from bottom to top;

2 When it is not particularly specified, the calibration of the bracing member should be carried out after the adjacent structure is calibrated and fixed;

3 The Buckling Restrained Brace shall be installed according to requirements of the design document and product instruction.

11.4.4 Installation of the truss (roof truss) shall be carried out after the calibration of the steel column is qualified; and it shall meet the following requirements:

1 Integral or section installation may be adopted for the steel truss (roof truss);

2 Steel truss (roof truss) shall be free from deformation in the process of pulling and hoisting;

3 During the installation of single steel truss (roof truss), a cable or rigid brace shall be adopted to increase the temporary lateral temporary restraint.

11.4.5 Installation of the steel plate shear wall shall meet the following requirements:

1 Measures to avoid any out-of-plane deformation shall be taken during the hoisting of the steel plate shear wall;

2 Installation time and sequence of the steel plate shear wall shall meet the requirements of the design document.

11.4.6 Installation of joint bearing nodes shall meet the following requirements:

1 Special tooling shall be adopted for the hoisting and installation of the joint bearing nodes;

2 The bearing assembly should not be disassembled for installation and temporary fixation measures shall be taken after the positioning;

3 During the assembling of the connecting pin and hole, they shall be fitted with each other; and tapered hole and axis should be adopted; special tools shall be adopted for tightening and installation;

4 After the installation, protection shall be made well for the finished product.

11.4.7 Installation of the steel casting or cast steel node shall meet the following requirements:

1 During the delivery, clear installation reference mark shall be provided;

2 The site welding and inspection shall be in strict accordance with the specific plan for the welding procedure.

11.4.8 During the hoisting of the heavy combined member that is assembled of multiple members on the ground, the positions and quantity of the hoisting points shall be determined through calculation.

11.4.9 Post-installation members shall be installed according to requirements of the design document or hoisting conditions and the processing length should be determined through the on-site actual measurement; when welding connection is adopted for the post-installation members and finished structure, measures to minimize the welding deformation and welding residual stress shall be taken.

11.5 Single-story Steel Structures

11.5.1 Hoisting of single-span structure should be carried out from one side of the end span to the other side, from the middle to the both ends or vice versa. For multi-span structure, the hoisting should be carried out from the major span to the minor span; when multiple hoisting equipment is available for joint operation, multiple spans also may be hoisted simultaneously.

11.5.2 In the process of installation, for the single-story steel structures, temporary braces or stable cables of column shall be timely installed and the installation shall be extended after it forms a stable space structure system, which shall be able to resist the structure self-weight, wind load, snow load, construction load and impact load in the process of hoisting.

11.6 Multi-story and Tall Steel Structures

11.6.1 Multiple assembly line sections should be divided for installation of the multi-story and tall steel structures with each section of framework as the unit. The division of the assembly line section shall meet the following requirements:

 1 The heaviest member within the assembly line section shall be within the hoisting capacity of the hoisting equipment;

 2 Climbing height of the hoisting equipment shall meet the requirements of the hoisting height of the member within the next assembly line section;

 3 Length of the column within each assembly line section shall be determined according to factors such as factory processing, transportation, stacking and site hoisting. 2~3 story heights should be taken for the length and the sectioning positions should be located at 1.0m~1.3m above the elevation of the beam top;

 4 Division of the assembly line sections shall be corresponding with the construction of the concrete structures;

 5 For each assembly line section, the assembly line area may be divided on the plane according to the structure characteristics and field conditions for construction.

11.6.2 Hoisting of the member within the assembly line sections should meet the following requirements:

 1 The sequence of first column and then beam within the whole assembly line section or that in local may be adopted for the hoisting. The single column shall not be at the cantilever for a long time;

 2 Installation of the steel floor slab and profiled metal plate shall be synchronous to the hoisting progress of the member;

 3 Hoisting sequence within special assembly line section shall be determined according to installation process and shall meet requirements of the design document.

11.6.3 Installation calibration of the multi-story and tall steel structures shall be carried out according to the reference column and shall meet the following requirements:

 1 The reference column shall be able to control the plane dimension of buildings and be convenient for calibration of other columns; corner column should be selected as a reference column;

 2 Appropriate measuring instrument and calibration tool should be adopted for the calibration of the steel column;

 3 After the calibration of the reference column is completed, the other columns shall be calibrated.

11.6.4 During the installation of the multi-story and tall steel structures, relative elevation or design elevation may be adopted to control the story elevation and shall meet the following requirements:

 1 When the design elevation is adopted for the control, each segmented column shall be regarded as a unit to adjust the column elevation and the elevation of each segmented column shall meet the design requirements;

 2 Allowable tolerance of the total building height or the height difference at the top of each segmented column in the same story shall meet the requirements of the current national standard GB 50205 *Code for Acceptance of Construction Quality of Steel Structures*.

11.6.5 For the segmented column in the same assembly line section and at the same installation height, the positioning axis of the next segmented column shall be led from the ground, after all the column members are installed, calibrated, connected and qualified in the inspection.

11.6.6 For the installation of tall steel structures, corresponding analysis is carried out for the effect of the vertical compressive deformation on the structure; and corresponding measures such as installation elevation presetting and post-connecting member arrangement shall be taken according to structure characteristics and the effect.

11.7 Long-span Spatial Steel Structures

11.7.1 Installation methods such as spare parts in air, strip/block hoisting, sliding, unit or integral lifting (jacking), integral hoisting, deployable integral lifting and high-level assembling may be adopted for the long-span spatial steel structures according to structure characteristics and site construction conditions.

11.7.2 Division of spatial structure hoisting units shall be determined according to factors such as structure characteristics, transportation mode, hoisting equipment performance and installation site conditions.

11.7.3 Construction of cable (prestress) structures shall meet the following requirements:

 1 Before construction, acceptance shall be carried out for the delivery reports, product warranties and inspection reports of the steel cable, anchorage, parts and accessories as well as length, diameter, variety, specification, color/luster and quantity of the cable; and the prestressed construction shall be carried out after it is qualified in the acceptance;

 2 Before tension for the construction of the cable (prestress) structure, whole process structure analysis of construction stage shall be carried out and the tension sequence shall be determined with the analysis results as the basis; specific construction plan for the cable (prestress) shall be prepared;

3 Before the tension for the construction of the cable (prestress) structure, section acceptance shall be carried out for the steel structure and prestress tension may be carried out only after it is qualified in the acceptance;

4 The cable (prestress) tension shall meet the principles of staging, grading, symmetrical, slow and uniform, and synchronous loading and the extra tension requirements shall be determined according to characteristics of the structure and materials;

5 Cable force and structural deformation monitoring should be carried out for the cable (prestress) structure and a monitoring report shall be prepared.

11.7.4 For the construction of the long-span spatial steel structures, the influence of the ambient temperature changes on the structure shall be analyzed.

11.8 High-rising Steel Structures

11.8.1 Installation methods such as spare parts (unit) in the air, integral pulling and lifting (jacking) may be adopted for the high-rising steel structures.

11.8.2 When integral pulling is adopted for the installation of the high-rising steel structures, the quantity and positions of the hoisting points shall be determined through calculation and structural safety checking shall be carried out under different construction inclination angles or conditions of the structures in the process of integral pulling.

11.8.3 When the elevation and axis reference points for the installation of high-rising steel structures pass up, influences of the wind load, ambient temperature and sunlight on the structural deformation shall be analyzed.

12 Profiled Metal Plate

12.0.1 This Chapter is applicable to construction of profiled metal plate for the composite slab in the story and platform and non-composite slab for permanent concreting form.

12.0.2 Before installation of the profiled metal plate, arrangement drawing of profiled metal plates in each story shall be drawn and the drawing shall cover the specification, dimension and quantity of the profiled metal plates; detailed drawing for the bearing structure and connection of major structure as well as the edge-sealing baffle.

12.0.3 Before the installation of the profiled metal plate, the position line of the profiled metal plate shall be marked on the bearing structure. When it is laid, the waveform groove at the profiled metal plate ends shall be aligned.

12.0.4 Special hoisting mechanism shall be adopted for handling and transportation of the profiled metal plate and the steel rope is strictly forbidden to be directly adopted for binding and hoisting.

12.0.5 Anchoring and bearing length of the profiled metal plate and major structure (steel beam) shall meet the design requirements and shall be greater than or equal to 50mm; spot welding, fillet welding or nail fastening may be adopted for the end anchorage and the layout position shall meet the design requirements.

12.0.6 The installation and connection of the profiled metal plate transferred to the floor shall be finished on the same day; if they are still surplus, they shall be fixed on the steel beam or transferred to the ground stockyard.

12.0.7 The surface of the steel beam bearing the profiled metal plate shall be maintained clean and the clearance between the profiled metal plate and the top surface of steel beam shall be controlled within 1mm.

12.0.8 When the sealing plate of the edge form is installed, it shall be aligned with the wave pitch of the profiled metal plate and the tolerance is less than or equal to 3mm.

12.0.9 The profiled metal plate shall be installed flat, smooth and straight and the plate surface shall be free from construction residues and contaminants.

12.0.10 When any equipment hole is required to be reserved on the profiled metal plate, plasma cutting or hollow drilling shall be used after the concreting, but the flame cutting shall not be adopted.

12.0.11 When the temporary support is required to be arranged at the construction stage according to the requirements of the design document, it shall be arranged before the concreting and may be removed only after the concrete reaches the specified strength. During the concreting, centralized loading shall be avoided on the profiled metal plate.

13 Coating

13.1 General Requirements

13.1.1 This Chapter is applicable to coating of anti-corrosive paint layer, anticorrosion of metal hot spraying, anticorrosion of hot dipping galvanizing, coating of fire-retardant coating layer.

13.1.2 Anticorrosive coating construction of the steel structure should be carried out after the inspection lots of member assembling/test assembling engineering are qualified in the construction quality acceptance. After the coating, member number should be marked on the member; weight, position of gravity center and positioning mark shall be marked on large members.

13.1.3 The coating of fire-retardant coating layer for the steel structure shall be carried out after the inspection lots of the installation engineering and anticorrosive coating engineering are qualified in the construction quality acceptance. When it is specified that the coating of anticorrosive coating layer may not be carried out for the member in the design document, the coating of fire-retardant coating layer may be directly carried out after it is qualified in the installation acceptance.

13.1.4 The construction process and technology of the anticorrosive coating engineering and fire-retardant coating engineering shall meet the requirements of this Code, design document, the coating product instruction as well as those of the national current relevant product standards.

13.1.5 Before the coating of anticorrosive coating layer, surface treatment shall be carried out for the steel materials according to requirements of this Code and design document. When it is not specified in the design document, suitable treatment methods may be adopted on the surface of the steel materials according to the requirements of the paint products.

13.1.6 Quality acceptance shall be carried out for the coating of anti-corrosive paint layer and coating of fire-retardant coating layer according to the relevant requirements of the current national standard GB 50205 *Code for Acceptance of Construction Quality of Steel Structures*.

13.1.7 Quality acceptance may be carried out for the anticorrosion of metal hot spraying and anticorrosion of hot dipping galvanizing according to the relevant requirements of the current national standards GB/T 9793 *Metallic and Other Inorganic Coatings-Thermal Spraying-Zinc, Aluminum and Their Alloys* and GB/T 11373 *The General Principle of Surface Preparation of Metallic Substrate for Thermal Spraying*.

13.1.8 Coating systems for the member surface shall be compatible with each other.

13.1.9 During the coating, corresponding environmental protection and labor protection measures shall be taken.

13.2 Surface Treatment

13.2.1 When the member adopts anticorrosive coating, the surface derusting grade may be in accordance with the relevant requirements of the design document and the current national standard GB 8923 *Rust Grades and Preparation Grades of Steel Surfaces before Application of Paints and Related Products*; mechanical derusting and manual derusting methods are adopted for the treatment.

13.2.2 Surface roughness of the member may be selected, on the basis of the different undercoats and derusting grades, according to those specified in Table 13.2.2; and it shall be in accordance with the relevant requirements of the current national standard GB/T 13288 *The Assessment of Profile Grades of Steel Surface before Application of Paint and Related Products-Comparator*.

Table 13.2.2 Surface Roughness of Members

Undercoat of steel materials	Derusting grade	Surface roughness Ra (μm)
Thermal spray zinc or aluminum	Grade Sa3	60~100
Inorganic zinc-rich	Grade Sa2½~Sa3	50~80
Epoxy zinc-rich	Grade Sa2½	30~75
Parts inconvenient for sand spraying	Grade St3	

13.2.3 The surfaces of the steel materials after treatment shall be free from welding slag, crater, dust, oil stain, water and burr. For the galvanized member, the surfaces of the steel materials after pickling and derusting shall show a metallic color and luster and be free from stains, rust stain and residual acid solution.

13.3 Coating of Anticorrosive Paint Layer

13.3.1 Brush method, manual roll painting, air spraying and high pressure airless spraying methods may be adopted for the coating of anticorrosive paint layer.

13.3.2 Ambient temperature and relative humidity during the coating for the steel structure shall meet not only requirements of the product instruction of paints, but also the following requirements:

  1 When the ambient temperature and relative humidity for coating are not specified in the product instruction, the ambient temperature should be 5℃~38℃ and the relative humidity shall be less than or equal to 85%. The surface temperature of the steel materials shall be 3℃ higher than the dew point temperature, but shall be less than or equal to 40℃.

  2 The surface of the object under construction shall be free from any condensation;

  3 In case of rain, fog, snow and strong wind, the open-air coating shall be stopped and the construction shall be avoided in the strong sunshine;

  4 Protection measures shall be taken within 4h after the coating to avoid any rain and dust;

  5 When the wind force is greater than Scale 5, spray coating should not be carried out outdoor.

13.3.3 During the preparation, the paints shall be mixed uniformly and used along with the mixture; and it shall not be randomly added with any diluent.

13.3.4 Suitable repaint interval time shall be provided for the construction between different coating layers and the maximum/minimum repaint interval time shall meet the requirements of the paints' product instruction. The construction shall be carried out after it exceeds the minimum repaint interval time while the construction shall be carried out according to the paints' product instruction when it exceeds the maximum repaint interval time.

13.3.5 The interval time of the surface derusting treatment and coating should be within 4h and it shall not exceed 12h when they are carried out in the workshop or on a fine day with relatively low humidity.

13.3.6 On both sides of the weld at the field welded parts, the area temporarily without coating should be set aside and shall meet those specified in Table 13.3.6. At the weld or on both sides of the weld, anti-corrosive paints without influence on the welding quality may also be coated.

Table 13.3.6 Weld Area Temporarily without Coating (mm)

Sketch	Thickness of steel plate, t	Width of area temporarily without coating, b
	$t<50$	50
	$50 \leqslant t \leqslant 90$	70
	$t>90$	100

13.3.7 Member repainting shall meet the following requirements:

 1 For the member whose surface is coated with the factory primer, in case of any re-rust or white zinc salt due to welding, flame calibration, insolation and scratch, it shall be repainted after the surface treatment according to the original coating requirements;

 2 In case of any coating damage and welding burn in the process of transportation and installation, it shall be repainted according to the original coating requirements.

13.4 Metal Hot Spraying

13.4.1 Air spraying or electric spraying may be adopted for the metal hot spraying of steel structures and it may be carried out according to the relevant requirements of the current national standard GB/T 9793 *Metallic and Other Inorganic Coatings-Thermal Spraying-Zinc, Aluminum and Their Alloys*.

13.4.2 Interval time of the steel structure surface treatment and hot spraying shall be within 12h on a fine day or climate conditions with less humidity; and it shall not exceed 2h on a rainy day, in humid environment or under climate conditions with salt mist.

13.4.3 Metal hot spraying shall meet the following requirements:

 1 The adopted compressed air shall be dry and clean;

 2 The spray gun should form a right angle with the surface and have uniform travel speed; the directions of the spray gun between spray coatings shall be vertical to each other and cross covered;

 3 Primary spraying thickness should be $25\mu m \sim 80\mu m$ and 1/3 overlapping width shall be provided between spraying belts in the same layer;

 4 When the ambient temperature is less than 5℃ or the surface temperature of the steel structure is 3℃ less than the dew point, hot spraying shall be stopped.

13.4.4 For the sealant for the metal hot spraying layer or the coating of the first pass of sealing paints, brush method should be adopted; and the construction process shall meet requirements of Section 13.3 in this Code.

13.5 Anticorrosion of Hot Dipping Galvanizing

13.5.1 Quality of hot dipping galvanizing per unit area of the member surface shall meet the requirements of the design document.

13.5.2 Hot dipping galvanizing of the member shall meet the relevant requirements of the current national standard GB/T 13912 *Metallic Coatings-Hot Dip Galvanized Coatings on Fabricated Iron and Steel Articles-Specifications and Test Methods* and measures against thermal deformation shall be taken.

13.5.3 Mechanical modes such as flattener, rolling or jack shall be taken for the rectification of the bending or distortion of member caused by hot dipping galvanizing. During the rectification, measures such as mat wood should be taken, but heating rectification shall not be adopted.

13.6 Coating of Fire-retardant Coating Layer

13.6.1 Before coating of fire-retardant coating layer, derusting and coating of anticorrosive coating layer for the surface of the steel materials shall meet the requirements of the design document and the current relevant standards of the nation.

13.6.2 The surface of the base layer shall be free from oil stain, dust and silt; and the anticorrosive coating layer shall be complete and the primer is free from missing brushing. The gap of the member joint shall be filled and leveled up with fire-retardant coating or other fire-retardant materials.

13.6.3 The selected fire-retardant coating shall meet the requirements of the design document and the current relevant standards of the nation; and it shall have impact resistance and adhesive strength, but not corrode the steel materials.

13.6.4 Fire-retardant coating may be stirred or prepared on the site according to requirements of the product instruction. The coating prepared on the same day shall be used up within the duration specified in the product instruction.

13.6.5 Under one of the following conditions, steel wire mesh connected with the member or other corresponding measures should be arranged in or taken for the thick fire-retardant coating layer:

  1 The steel beam bearing impact and vibration loads;
  2 The steel beam and truss with coating layer thickness no less than 40 mm;
  3 The member with the adhesive strength of the coating less than or equal to 0.05MPa;
  4 The steel plate wall and steel beam with the web height greater than 1.5m.

13.6.6 Methods such as spray coating, brush coating or roll coating may be adopted for the coating of fire-retardant coating layer.

13.6.7 The coating of fire-retardant coating layer shall be carried out in layers; after the previous coating is dry or cured, the construction for the next coating is carried out.

13.6.8 Under one of the following conditions, respray or repair coating shall be carried out for the thick fire-retardant coating layer:

  1 The coating layer is poorly dried and cured without firm adhesion or present with pulverization and shedding;
  2 The coating layer at the steel structure joint and corners is present with appreciable hollowness;
  3 The thickness of the coating layer is less than 85% of that specified in the design;
  4 The thickness of the coating layer fails to reach that specified in the design and the

continuous length of the coating layer is greater than 1m.

13. 6. 9 Construction of thin fire-retardant coating surface layer shall meet the following requirements:

1 The surface layer shall be coated after the undercoat is dry;

2 Coating of the surface layer shall be uniform and consistent in the color and the connecting raft shall be flat and smooth.

14 Construction Survey

14.1 General Requirements

14.1.1 This Chapter is applicable to the plane control, elevation control and detailed survey of steel structures.

14.1.2 Before the construction survey, the specific survey plan shall be prepared according to the requirements of design drawing and the installation requirements of steel structures.

14.1.3 Before the installation of steel structures, the construction control network shall be arranged.

14.2 Plan Control Network Survey

14.2.1 The plan control network may be arranged as the cross axis or rectangular control network according to site topographic condition and building structure form. Polygon control network may be arranged for the building with abnormal shape according to the shape of the building.

14.2.2 The axis control pile for the building shall be measured according to the plan control network; and methods such as rectangular coordinate, polar coordinate, angle (direction) intersection and distance intersection may be selected for the positioning and setting out.

14.2.3 For the plan control network of the building, the external control method should be adopted for the fourth story below and internal control method for the fourth story or above. The plan control network for the upper stories shall be based on the control network for the bottom story and vertically transferred one by one with an apparatus. One turning point should be arranged every 50m~80m for the vertical transfer, and the allowable error of the vertical transfer at the control point shall meet those specified in Table 14.2.3.

Table 14.2.3 Allowable Error of Vertical Transfer at the Control Point (mm)

Item		Allowable error for survey
Each story		3
Total height H	$H \leqslant 30m$	5
	$30m < H \leqslant 60m$	8
	$60m < H \leqslant 90m$	13
	$90m < H \leqslant 150m$	18
	$H > 150m$	20

14.2.4 The balancing check of control network shall be carried out after the axis control reference point is transferred to the middle construction stories. The adjusted point accuracy shall meet the requirements that the relative error of the side length shall reach 1/20000 and the relevant mean square error of the angle $\pm 10''$. When it is particularly specified in the design, the setting out accuracy shall be determined according to the tolerance.

14.3 Elevation Control Network Survey

14.3.1 The first grade elevation control network survey shall be arranged according to closed loop, annexed line or node network. The accuracy of elevation survey should not be less than the third-grade level accuracy.

14.3.2 The benchmarks of the elevation control points for the steel structures may be arranged either on the stake of the plan control network or the peripheral fixed ground object; it may also be separately embedded. The quantity of the benchmarks shall be greater than or equal to three.

14.3.3 The transfer of the building elevation should be carried out by the survey method through hanging a steel ruler. The temperature, ruler length and tension correction shall be carried out during the reading of the steel ruler. When the elevation is transferred up, it should be respectively transferred from two places. For the relatively large area or tall structures, it should be respectively transferred from three places. When the transferred elevation error is not greater than ±3.0mm, its average value may be regarded as the elevation reference of the construction stories. Otherwise, it shall be transferred again. The allowable error for survey of vertical-transfer elevation shall meet those specified in Table 14.3.3.

Table 14.3.3 Allowable Error for Survey of Vertical-transfer Elevation(mm)

Item		Allowable error for survey
Each story		±3
Total height H	$H \leqslant 30m$	±5
	$30m < H \leqslant 60m$	±10
	$H > 60m$	±12

Note: the error in this table excludes the deformation value caused by sedimentation and compression.

14.4 Survey of Single-story Steel Structures

14.4.1 Before the installation of steel column, center lines or installation lines shall be respectively drawn on the four sides of column, and the permissible error of the snapline is 1mm.

14.4.2 A theodolite shall be adopted on the direction of two mutually-perpendicular axes during the installation of the vertical steel column and the perpendicularity of the steel column is calibrated at the same time. When the observation side is not a constant section, the theodolite shall be installed on the axis; when the observation side is a constant section, the horizontal angle between the theodolite center line and axis shall not be greater than 15°.

14.4.3 The survey of crane beam in the steel structure building and rail installation shall meet the following requirements:

1 The center line of the crane beam shall be measured by the parallel handrail method according to the plan control network of the building; the allowable error for the transfer of the crane beam center line is ±3mm, and the allowable tolerance for the elevation of the base plate on the beam surface is ±2mm;

2 For the upper crane beam rail, the allowable error for the transfer of its center line is ±2mm, and the spacing between the intermediate densification control points shall not be greater than twice of the column spacing. All points shall be parallelly measured to the side of the top

bracket near the column and be regarded as the basis of the rail installation.

3 A level shall be erected on the column bracket surface and the rail installation elevation is determined according to third-grade level precision. The allowable error of the elevation control point is ±2mm, that of the rail span is ±2mm, that for the transfer of the rail center line is ±2mm and that of the rail elevation point is ±1mm.

14.4.4 The actual survey records such as perpendicularity, straightness, elevation and deflection (arch camber) shall be provided after the installation of the steel roof truss (truss).

14.4.5 The positioning of the complex member may either be determined with three-dimensional coordinate by the total station which is directly erected on the control point, or be jointly measured and controlled by the water level (for elevation) and total station (for plane coordinate).

14.5 Survey of Multi-story and Tall Steel Structures

14.5.1 Before the installation of multi-story and tall steel structures, the building positioning axis, the bottom-story column axis and the column-bottom foundation elevation shall be rechecked. The installation shall be started after the rechecking is deemed as acceptable.

14.5.2 The control axis of each segment of steel column shall be measured from the turning point of the datum control axis and shall not be led from the axis of the column on the lower story.

14.5.3 Before the installation of the steel beam, the perpendicularity change of the columns at both ends of the steel beam shall be measured and also the perpendicularity change due to the beam connection of adjacent columns shall be monitored. After completion of the integral member installation in certain area, the integral structure re-survey shall be carried out.

14.5.4 During the installation of the steel structures, the possible member expansion or bending deformation caused by factors such as sunlight and welding shall be analyzed and corresponding measures shall be taken. During the installation, the following items should be observed and recorded:

1 The perpendicularity error of the column caused by the weld shrinkage of columns and beams;

2 The deformation of the steel columns under the influence of the sunlight temperature difference and wind force;

3 The influence of tower crane (attachment or ascending) on the structure perpendicularity.

14.5.5 The allowable tolerance for the integral perpendicularity of major structure is $H/2500+10mm$ (H is height), and shall be less than or equal to 50.0mm; the allowable tolerance of integral plane bending is $L/1500$ (L is width), and shall be less than or equal to 25.0mm.

14.5.6 For over 150m high building steel structures, GPS or corresponding method should be adopted to recheck the integral perpendicularity.

14.6 Survey of High-rising Steel Structures

14.6.1 The construction control network of the high-rising steel structures should be arranged as farmland, round or radial patterns on the ground.

14.6.2 When the construction axis point is directly measured by transferring the plan control point to the upper part, different survey methods shall be adopted for checking and the allowable error for survey is 4mm.

14.6.3 The laser plummet apparatus should be used for the survey of tower verticality (the elevation is above ±0.000m) and the diameter of the track circle at the laser point which is drawn by the 360° rotation of the laser apparatus received by the receiving target at 100m elevation shall be less than 10mm.

14.6.4 The laser plummet apparatus should be arranged at the tower center point when the elevation of the high-rising steel structures is less than 100m, four apparatus should be arranged when the elevation is 100m~200m, and five apparatus (including the tower center point) should be arranged when the elevation is above 200m. The point position of the apparatus shall be directly measured from the tower axis point, and checked by different survey methods.

14.6.5 The permissible error for survey with the laser plummet apparatus transferred to the receiving target shall meet those specified in Table 14.6.5. For the high-rising steel structures with special requirements, their allowable error shall be jointly determined by the design and construction organizations.

Table 14.6.5 Allowable Error for Survey with the Laser Plummet Apparatus Transferred to the Receiving Target

Tower height (m)	50	100	150	200	250	300	350
The allowable tolerance of high-rising structures acceptance (mm)	57	85	110	127	143	165	—
The allowable error for survey (mm)	10	15	20	25	30	35	40

14.6.6 When the construction of the high-rising steel structures reaches 100m high, the sunlight deformation observation should be carried out; in addition, the sunlight deformation curve is drawn and the minimum sunlight deformation range is listed.

14.6.7 Round-trip survey along the tower direction should be carried out to determine the elevation of the high-rising steel structures with a steel ruler; the survey results should be corrected with regard to ruler length, temperature & tension and their precision shall be greater than 1/10000.

14.6.8 GPS should be adopted to survey and recheck the integral perpendicularity of over 150m high high-rising steel structures.

15 Construction Monitoring

15.1 General Requirements

15.1.1 This Chapter is applicable to the construction monitoring for the large and important steel structures, such as tall structures, long-span spatial structures and high-rising structures according to the design requirements and the contract provisions.

15.1.2 The construction monitoring methods shall be selected according to actual engineering situations such as monitoring objects, monitoring purpose, monitoring frequency, monitoring duration and monitoring precision.

15.1.3 During the construction of steel structures, the process monitoring may be carried out for items such as structural deformation, structural internal force and environmental capacity. The specific monitoring items and parts of the steel structures may be selected according to different engineering requirements and construction conditions.

15.1.4 The adopted monitoring instrument and equipment shall meet the data accuracy requirement and guarantee the data stability and accuracy; the sensors with high sensitivity, good corrosion resistance, strong electromagnetic wave interference resistance, small volume and light weight should be adopted.

15.2 Construction Monitoring

15.2.1 The special construction monitoring plan shall be prepared for the construction monitoring.

15.2.2 Reliable protection measures shall be taken for the arrangement of construction monitoring points according to the site installation conditions and cross construction operation. The stress detector shall be arranged at the part with the most unfavorable structure stress or the feature part according to the design requirements and working demands. The deformation sensor or measuring point should be arranged at the part with the relatively large structural deformation. And the temperature sensor should be arranged at the structural feature section and uniformly distributed along four sides and the elevation.

15.2.3 The grading and precision requirement of deformation monitoring for steel structures shall meet those specified in Table 15.2.3.

15.2.4 The deformation monitoring method may be selected according to those specified in Table 15.2.4 or various methods may be simultaneously adopted for the monitoring. The stress strain should be monitored with the sensors such as stressometer and strainometer.

15.2.5 Monitoring data shall be collected and processed in time, particularly in accordance with the frequency. The missed, error or abnormal data shall be monitored, ratified or modified.

15.2.6 The monitoring period of stress and strain should be synchronous with that of deformation.

15.2.7 When the structural deformation and structural internal force are monitored, the environmental variables such as temperature and wind force at the monitoring points should be monitored at the same time.

Table 15.2.3 The Grading and Precision Requirement of Deformation Monitoring for Steel Structures

Grades	Vertical displacement monitoring		Horizontal displacement monitoring	Application scope
	Mean square error of height at deformation observation point (mm)	Mean square error of height difference at adjacent deformation observation point (mm)	Mean square error of a point at deformation observation point (mm)	
Grade I	0.3	0.1	1.5	Tall buildings, special structures, high-rising buildings and industrial buildings which are particularly easily subject to deformation
Grade II	0.5	0.3	3.0	Tall buildings, special structures, high-rising buildings and industrial buildings which are relatively easily subject to deformation
Grade III	1.0	0.5	6.0	Common tall buildings, special structures, high-rising buildings and industrial buildings

Notes: 1 The mean square error of height and mean square error of a point at the deformation observation point refer to the mean square error relative to the adjacent datum;
2 $1/\sqrt{2}$ of corresponding mean square error of a point in this table may be taken as the limit of the mean square error of displacement in particular direction.
3 For the vertical displacement monitoring, the monitoring precision grades may be determined according to mean square error of height at deformation observation point or mean square error of height difference at adjacent deformation observation point.

Table 15.2.4 The Selection of Deformation Monitoring Method

Types	Monitoring Method
Horizontal deformation monitoring	Triangular network, polar coordinate, intersection, GPS measurement, normal and inverted vertical line, collimation line, tension wire alignment, laser alignment, precise distance measurement, extensometer, multi-point displacement and inclinometer methods
Vertical deformation monitoring	Leveling, hydro-static leveling, electromagnetic distance measurement and trigonometric leveling methods
Three-dimensional displacement monitoring	Auto tracking total station survey and real-time satellite positioning measurement methods
Declivity of main body building	Transit projection, differential settlement, laser alignment, normal line, inclinometer, electrical vertical beam methods
Deflection observation	Normal line, differential settlement, displacement meter and deflectometer methods

15.2.8 The quantitative analysis and qualitative analysis shall be timely carried out for the monitoring data. Methods such as chart analysis, statistical analysis, comparative analysis and modeling analysis may be adopted for the analysis of the monitoring data.

15.2.9 When the monitoring results are required to be used for the tendency forecast, the error range and applicable conditions of the forecast result shall be given.

16　Construction Safety and Environmental Protection

16.1　General Requirements

16.1.1　This Chapter is applicable to the construction safety and environmental protection of the steel structures.

16.1.2　Before the construction of steel structures, the special construction safety & environmental protection plan and safety emergency plan shall be prepared.

16.1.3　The safety production education and training shall be carried out for operation personnel.

16.1.4　New operation personnel shall receive the three-level safety education. In case of any work type change, the operation personnel shall firstly receive the training on operating skill and safety operation knowledge. Those who fail in the safety production education and training shall not take up their posts.

16.1.5　During the construction, the qualified labor protection articles in accordance with the requirements of the current relevant standards of the nation shall be provided for the operation personnel; and the operation personnel shall be trained and supervised to use them correctly.

16.1.6　For the operation easy to cause an occupational disease, the special protection measures shall be taken for the operation personnel.

16.1.7　The high-altitude operation must not be carried out when all safety measures are unqualified in the inspection.

16.2　Climb Up

16.2.1　Scaffold erection shall meet the relevant requirements of the current professional standards JGJ 130 *Technical Code for Safety of Steel Tubular Scaffold with Couplers in Construction* and JGJ 166 *Technical Code for Safety of Cuplok Steel Tubular Scaffolding in Construction*. When other climb-up measures are adopted, the structural safety calculation shall be carried out.

16.2.2　The person-and-cargo elevator shall be adopted for the climbing up during the construction of multi-story and tall steel structures. And reasonable safety climb-up facilities shall be erected for the stories inaccessible by the elevator.

16.2.3　When the hook for the hoisting of the steel column loosens, construction personnel should climb up through a steel hook ladder and shall adopt a safety catcher to protect themselves. The steel hook ladder shall be firmly connected with steel column in advance and be lifted with the column.

16.3　Safety Channel

16.3.1　Plane safety channel required for the installation of steel structures shall be continuously erected story by story.

16.3.2　The plane safety channel width for the construction of steel structures should be greater than or equal to 600mm, and both sides shall be arranged with safety guardrail or protective wire rope.

16.3.3 The operation personnel walking on the steel beam or steel truss shall wear double hook safety belt.

16.4 Protection of Portals and Sides

16.4.1 The rigid cover plate shall be adopted for fixation and protection of the portal whose side length or diameter is 20cm~40cm, the steel pipe scaffold shall be erected for the portal whose side length or diameter is 40cm~150cm, and the dense mesh safety network protection and guardrail shall be erected for the portal whose side length or diameter is above 150cm.

16.4.2 After hoisting of the steel beam in building stories, the safety net shall be timely laid zone by zone.

16.4.3 After hoisting of the steel beam around the story, the guardrail shall be arranged near the edge of each story, and its height shall not be less than 1.2m.

16.4.4 The edge scaffolds, operation platforms, and safety nets shall be reliably fixed on the structures during their erection.

16.5 Construction Machinery and Equipment

16.5.1 All construction machineries used for the construction of steel structures shall meet the relevant requirements of the current professional standard JGJ 33 *Technical Specification for Safety Operation of Construction Machinery*.

16.5.2 The hoisting machinery shall be equipped with limiting device and regularly inspected.

16.5.3 The special technical plan shall be provided during the installation and dismantling of tower cranes.

16.5.4 The measures to prevent tower cranes from mutual impact shall be adopted for the operation of group towers.

16.5.5 Good grounding device shall be provided for the tower cranes.

16.5.6 When the hoisting machinery (non-approved product) is adopted, the design calculation must be conducted and the safety checking shall be conducted.

16.6 Safety of Site Hoisting Area

16.6.1 The security cordon shall be arranged around the hoisting area and non-operation personnel must not enter into it.

16.6.2 When the hoisting object is hoisted 200mm~300mm away from the ground, the overall inspection shall be carried out. The object shall be hoisted again after it is confirmed without any error.

16.6.3 When the air speed reaches 10m/s, the hoisting operation should be stopped and when the air speed reaches 15m/s, the hoisting operation shall not be carried out.

16.6.4 Anti-falling measures shall be taken for the small hand tools and small parts used for the high-altitude operation.

16.6.5 The construction power supply shall meet the requirements of the current professional standard JGJ 46 *Technical Code for Safety of Temporary Electrification on Construction Site*.

16.6.6 Professional personnel shall be assigned to be responsible for the installation, maintenance and management of electric equipment and electric wire at the construction site.

16.6.7 Before the installation of members hoisted to the story or the roof is finished, the reliable temporary fixation measures shall be taken.

16.6.8 In case of any water, ice, frost or snow on the surface of the profiled steel plate, it shall be cleaned up in time and the relevant anti-skidding measures shall be taken.

16.7 Fire Safety Measures

16.7.1 The relevant fire safety management system shall be provided before the construction of steel structures.

16.7.2 The use of fire in the site construction operation shall be approved by the relevant departments.

16.7.3 Fire protection facilities and evacuation facilities shall be arranged at the construction site, and the fire patrolling shall be organized regularly.

16.7.4 During the gas cutting and high-altitude welding, the hazardous substances and combustibles in the operating area shall be cleaned up and fire measures shall be taken.

16.7.5 During the construction of site painting and fire-retardant coating, the product storage and fire protection shall be in accordance with the requirements of product instruction.

16.8 Environmental Protection Measures

16.8.1 During the construction, the noise shall be controlled, the construction time shall be arranged properly, and the influence on the surroundings shall be minimized.

16.8.2 The construction area shall be maintained clean.

16.8.3 Night construction light shall be illuminated toward the inside site and protection measures shall be taken for the welding arc.

16.8.4 The declaration formalities shall be well conducted for the night construction and the construction shall be in accordance with the requirements of relevant government departments.

16.8.5 During the construction of site painting and fire-retardant coating, anti-pollution measures shall be taken.

16.8.6 The waste and leftover materials at the installation site of steel structures shall be properly collected by classification and be uniformly recovered and recycled, and shall not be disposed and stacked randomly.

Explanation of Wording in This Code

1 Words used for different degrees of strictness are explained as follows in order to mark the differences in executing the requirements in this Code:

　　1) Words denoting a very strict or mandatory requirement:
　　　"Must" is used for affirmation; "must not" for negation.

　　2) Words denoting a strict requirement under normal conditions:
　　　"Shall" is used for affirmation; "shall not" for negation.

　　3) Words denoting a permission of a slight choice or an indication of the most suitable choice when conditions permit:
　　　"Should" is used for affirmation; "should not" for negation.

　　4) "May" is used to express the option available, sometimes with the conditional permit.

2 "Shall comply with..." or "Shall meet the requirements of..." is used in this Code to indicate that it is necessary to comply with the requirements stipulated in other relative standards.

List of Quoted Standards

1. GB 50009 *Load Code for the Design of Building Structures*
2. GB 50017 *Code for Design of Steel Structures*
3. GB/T 50105 *Standard for Structural Drawings*
4. GB 50135 *Code for Design of High-rising Structures*
5. GB 50205 *Code for Acceptance of Construction Quality of Steel Structures*
6. GB 50300 *Unified Standard for Constructional Quality Acceptance of Building Engineering*
7. GB 50661 *Code for Welding of Steel Structures*
8. GB/T 196 *General Purpose Metric Screw Threads-Basic Dimensions*
9. GB/T 197 *General Purpose Metric Screw Threads-Errors*
10. GB/T 222 *Permissible Errors for Chemical Composition of Steel Products*
11. GB/T 223 *Methods for Chemical Analysis of Iron, Steel and Alloy*
12. GB/T 228.1 *Metallic materials-Tensile testing-Part 1: Method of Test at Room Temperature*
13. GB/T 229 *Metallic materials-Charpy Pendulum Impact Test Method*
14. GB/T 232 *Metallic Materials-Bend Test*
15. GB/T 247 *General Rule of Package Mark and Certification for Steel Plates (Sheets) and Strips*
16. GB/T 324 *Weld Symbolic Representation on Drawings*
17. GB/T 699 *Quality Carbon Structural Steels*
18. GB/T 700 *Carbon Structural Steels*
19. GB/T 706 *Hot Rolled Section Steel*
20. GB/T 708 *Dimension Shape Weight and Error for Cold-rolled Steel Plates and Sheets*
21. GB/T 709 *Dimension Shape Weight and Error for Hot-rolled Steel Plates and Sheets*
22. GB/T 882 *Clevis Pin with Head*
23. GB 912 *Hot-rolled Sheets and Strips of Carbon Structural Steels and High Strength Low Alloy Structural Steels*
24. GB/T 1228 *High Strength Bolts with Large Hexagon Head for Steel Structures*
25. GB/T 1229 *High Strength Large Hexagon Nuts for Steel Structures*
26. GB/T 1230 *High Strength Plain Washers for Steel Structures*
27. GB/T 1231 *Specifications of High Strength Bolts with Large Hexagon Head, Large Hexagon Nuts, Plain Washers for Steel Structures*
28. GB/T 1591 *High Strength Low Alloy Structural Steels*
29. GB/T 2101 *General Requirement of Acceptance Packaging, Marking and Certification for Section Steel*
30. GB/T 2975 *Steel and Steel Products-Location and Preparation of Test Pieces for Mechanical Testing*
31. GB/T 3077 *Alloy Structure Steels*
32. GB/T 3098.1 *Mechanical Properties of Fasteners-Bolts, Screws and Studs*

33　GB/T 3274　*Hot-rolled Plates and Strips of Carbon Structural Steels and High Strength Low Alloy Structural Steels*

34　GB/T 3632　*Sets of Torshear Type High Strength Bolt Hexagon Nut and Plain Washer for Steel Structures*

35　GB/T 4171　*Atmospheric Corrosion Resisting Structural Steel*

36　GB/T 4842　*Argon*

37　GB/T 5117　*Carbon Steel Covered Electrodes*

38　GB/T 5118　*Low Alloy Steel Covered Electrodes*

39　GB/T 5224　*Steel Strand for Prestressed Concrete*

40　GB/T 5293　*Carbon Steel Electrodes and Fluxes for Submerged Arc Welding*

41　GB/T 5313　*Steel Plates with Through-thickness Characteristics*

42　GB/T 5780　*Hexagon Head Bolts-Product Grade C*

43　GB/T 5781　*Hexagon Head Bolts-Full Thread-Product Grade C*

44　GB/T 5782　*Hexagon Head Bolts*

45　GB/T 5783　*Hexagon Head Bolts-Full Thread*

46　GB/T 5796　*Trapezoidal Screw Threads*

47　GB/T 6052　*Industrial Liquid Carbon Dioxide*

48　GB/T 6728　*Cold Formed Steel Hollow Sections for General Structure-Dimensions, Shapes, Weight and Permissible Deviations*

49　GB 6819　*Dissolved Acetylene*

50　GB/T 7659　*Steel Casting Suitable for Welded Structure*

51　GB/T 8110　*Welding Electrodes and Rods for Gas Shielding Arc Welding of Carbon and Low Alloy Steel*

52　GB/T 8162　*Seamless Steel Tubes for Structural Purposes*

53　GB 8918　*Steel Wire Ropes for Important Purposes*

54　GB 8923　*Rust Grades and Preparation Grades of Steel Surfaces Before Application of Paints and Related Products*

55　GB/T 9793　*Metallic and Other Inorganic Coatings-Thermal Spraying-Zinc, Aluminium and Their Alloys*

56　GB/T 10045　*Carbon Steel Flux Cored Electrodes for Arc Welding*

57　GB/T 10432.1　*Unthreaded Studs for Drawn Arc Stud Welding with Ceramic Ferrule*

58　GB/T 10433　*Cheese Head Studs for Arc Stud Welding*

59　GB/T 11263　*Hot-rolled H and Cut T Section Steel*

60　GB/T 11352　*Carbon Steel Castings for General Engineering Purpose*

61　GB/T 11373　*The General Principle of Surface Preparation of Metallic Substrate for Thermal Spraying*

62　GB/T 12470　*Low-alloy Steel Electrodes and Fluxes for Submerged Arc Welding*

63　GB/T 12755　*Profiled Steel Sheet for Building*

64　GB/T 13097　*Epichlorohydrin for Industrial Use*

65　GB/T 13288　*The Assessment of Profile Grades of Steel Surface before Application of Paint and Related Products-Comparator*

66　GB/T 13793　*Steel Pipes with a Longitudinal Electric (Resistance) Weld*

67 GB/T 13912 *Metallic Coatings-Hot Dip Galvanized Coatings on Fabricated Iron and Steel Articles-Specifications and Test Methods*

68 GB/T 14370 *Anchorage, Grip and Coupler for Prestressing Tendons*

69 GB 14907 *Fire Resistive Coating for Steel Structure*

70 GB/T 14957 *Steel Wires for Melt Welding*

71 GB/T 14977 *General Requirement for Surface Condition of Hot-rolled Steel Plates*

72 GB 16912 *Safety Technical Regulation for Oxygen and Relative Gases Produced with Cryogenic Method*

73 GB/T 17101 *Hot-dip Galvanized Steel Wires for Bridge Cables*

74 GB/T 17395 *Dimensions, Shapes, Masses and Errors of Seamless Steel Tubes*

75 GB/T 17493 *Low Alloy Steel Flux Cored Electrodes for Arc Welding*

76 GB/T 17505 *Steel and Steel Products General Technical Delivery Requirements*

77 GB/T 19879 *Steel Plates for Building Structure*

78 GB/T 20066 *Steel and Iron-Sampling and Preparation of Samples for the Determination of Chemical Composition*

79 GB/T 20934 *Steel Tie Rod*

80 JGJ 33 *Technical Specification for Safety Operation of Construction Machinery*

81 JGJ 46 *Technical Code for Safety of Temporary Electrification on Construction Site*

82 JGJ 85 *Technical Specification for Application of Anchorage, Grip and Coupler for Prestressing Tendons*

83 JGJ 130 *Technical Code for Safety of Steel Tubular Scaffold with Couplers in Construction*

84 JGJ 166 *Technical Code for Safety of Cuplok Steel Tubular Scaffolding in Construction*

85 YB/T 152 *High Strength Low Relaxation Hot-dip Galvanized Steel Strand for Prestress*

86 YB 3301 *The Welded Steel H-section*

87 YB/T 5004 *Zinc-coated Steel Wire Strands*

88 JB/T 6046 *Welding Assembly for Carbon Steel and Low Alloy Steel Post-welding Heat Treatment Method*

89 JB/T 10375 *Recommended Practice for Vibration Stress Relief on Welding Structure*

90 HG/T 2537 *Carbon Dioxide for Welding Use*

91 HG/T 3661.1 *Burning Gases for Welding and Cutting-Propene*

92 HG/T 3661.2 *Burning Gases for Welding and Cutting-Propane*

93 HG/T 3668 *Zinc Rich Primer*

94 HG/T 3728 *Mixed Gas for Welding-Argon-Carbon Dioxide*